European Commission

WEGA II Large wind turbines
Intermediate design report

WEGA II Large wind turbines
Scientific evaluation project

Contract No JOU-CT93-0349

Directorate-General XII
Science, Research and Development

1996

EUR 16902 EN

Published by the
European Commission

**Directorate-General
Telecommunications, Information Market and Exploitation of Research**

Bâtiment Jean Monnet
L-2920 Luxembourg

LEGAL NOTICE

Neither the European Commission nor any other person acting on behalf of the
Commission is responsible for the use which might be made of the following information.

The report is based on work done in an R&D project in the framework of an ongoing programme
of the European Commission. This programme, called JOULE,
supports research and development into renewable energy technologies.
It is managed by Directorate-General XII for Science, Research and Development.
The report, as published in this book, is supported by the Commission under EC Contract JOU2-CT93-0349.

This report has been compiled and edited by
E. Hau, G. Barton, J. Langenbrinck, Etaplan GmbH, Munich (Germany)
under subcontract of Elsamprojekt A/S Fredericia (Denmark)

A great deal of additional information on the European Union is available on the Internet.
It can be accessed through the Europa server (http://europa.eu.int)

Cataloguing data can be found at the end of this publication

Luxembourg: Office for Official Publications of the European Communities, 1996

ISBN 92-827-8903-9
© ECSC-EC-EAEC, Brussels • Luxembourg, 1996

Printed in Italy

Preface

European wind energy technology and utilisation has made enormous progress over the past decade. At the end of 1995 the installed wind power in the European Community totalled about 2,000MW. For large scale implementation in the future, it may be useful to focus on larger wind turbines in the megawatt class. One major reason is the problem of finding enough environmentally acceptable erection sites.

The European Commission has supported the development and demonstration of large wind turbines for more than ten years. The latest achievements in technology and economics are concentrated in the WEGA-II Large Wind Turbine Programme within the framework of the JOULE II programme. While the wind turbines of the former WEGA-I programme were merely experimental, this second generation considers the needs of the market and involves the major industry in the field.

From WEGA-I to WEGA-II an enormous progress has been made in reducing the machines' weight and costs and thus improving the economics. This has considerably motivated the European wind turbine manufacturers and ensures their leading position on the world market and the continuing development of wind power as a virtual supply option for the European energy supply industry.

At the end of 1995 the prototypes of the large wind turbines developed in the frame of the WEGA-II programme have been erected. This report describes the design of the turbines and also includes some first operational results as far as they were available in April 1996.

Komninos Diamantaras

Content

		page
1	**Introduction**	1
2	**Research and Development Projects within the Joule Programme**	7
	2.1 Bonus 750kW	9
	2.2 Enercon E-66	25
	2.3 Nordic 1000	37
	2.4 Vestas V63	47
	2.5 WEG MS4-600kW	57
3	**Demonstration Projects supported by THERMIE evaluated within the WEGA II Programme**	65
	3.1 AEOLUS II and Näsudden II	67
	3.2 West GAMMA 60	83
	3.3 NedWind NW53/NW55	95
	3.4 Nordtank NTK1500	107
4	**Scientific Measurement and Evaluation Project (SMEP)**	119
5	**Further European Wind Turbines in the Megawatt Class**	127
	5.1 HSW 1000	129
	5.2 Nordex N52	137
	5.3 Tacke TW 1.5	147

References

1
Introduction

Large experimental wind turbines in the megawatt class have now been built and tested for about twenty years. The first generation of multi-megawatt turbines built in the USA, Sweden and Germany proved to be useful in terms of wind energy technology R&D, but failed completely to meet the commercial requirements.

A second generation of large wind turbines has been built in the late eighties. They have been somewhat smaller in size, but they neither could achieve a commercial status.

All projects of this kind required a bigger R&D effort than expected at the beginning of the development. There have also been a number of technical problems and failures. However, the main obstacle to commercial use were the high investment costs and therefore bad economics.

For almost a decade, the European Commission has supported the research and development and in some cases the demonstration of large wind turbines in Europe within binational or multinational programmes. Projects focusing on R&D have been supported within the JOULE programme, other projects having already passed the R&D phase are demonstrated within the framework of the THERMIE programme.

The JOULE programme comprised a special programme for large wind turbines called WEGA (I) (German: "Wind Energie Große Anlagen"). It included three large wind turbines /1/.

Although the WEGA-I programme had not yet ended, one important conclusion could already be drawn from the first phase of the testing of the turbines. Large wind turbines in the megawatt class could not become commercially viable by using the design on which the WEGA-I machines have been based. The main reason is the high weight of the components (towerhead weight) which has led to extremely high manufacturing costs /2/.

In order to address the problem, the Commission (DG XII) has initiated a strategic study on large wind turbine technology and economics /3/. The study was based on the knowledge gathered from the large experimental WECs and on the experience gained from the commercial turbines which gradually increased up to 500kW in size.

The encouraging results of the study together with the progress in design caused the Commission to support a new generation of large wind turbines. This resulted in the WEGA-II programme which has been initiated by the Commission within the framework of the JOULE programme 1990 to 1994.

At the beginning of 1991, the Commission adressed the commercial manufacturers of wind turbines by a "Call for Manifestation of Interest in the Field of Large Wind Turbines". The technical document, which was handed to industrial manufacturers, included the preferred size of wind turbines, essential criteria for the evaluation of the design and the economic potential.

Major European manufacturers reacted to this call by proposing ten wind turbine projects in the range of 750 to 1,500kW with a rotor diameter of 50 to 66m. After thorough evaluation by a group of independent experts on wind energy, the Commission supported five projects in the framework of the JOULE programme, called WEGA-II. Further projects were supported within the THERMIE programme.

An essential point of the WEGA-II programme was the establishment of an independent "Scientific Measurement and Evaluation Project". This project includes among others the wind turbine projects supported in the JOULE and the THERMIE programme. The coordinator is ELSAMPROJEKT A/S (Denmark), moreover, numerous sub-contractors and wind energy experts are involved /4/.

The report describes the design of large wind turbines developed or evaluated within the WEGA-II programme. As far as already available, the first operational results are also included. It should be noted that the design features and the technical data refer to the first prototypes. The later machines produced in series will not necessarily correspond exactly to these prototypes.

The final results of the WEGA-II programme will be resumed in a "Synopsis Report" to be edited after a 12-month testing phase of the wind turbines.

Commission of the European Community - DG XII
Programme JOULE, R&D in the field of non-nuclear energies
and rational use of energy

Call for Manifestations of Interest in the Field of Large Wind Turbines
(Copy of the original call in 1991)

In view of large-scale utilisation of wind energy in electricity generation networks large turbines in the 1MW range would be particularly suitable. Pilot plants in this power range developed until now in the European Community could not yet demonstrate economic viability of this option. Recent assessment work provides evidence, however, that there are good prospects for further technological development of the large wind turbine option.

The aim of this section of the JOULE programme is to support development of a new generation of MW or near MW rated wind turbine generators of designs with sufficient potential to become competitive in the short term with the best medium size wind turbines (200 - 500kW).

In order to identify projects suitable for financial support, the Commission:

1. Calls for manifestations of interest in projects of design, construction and testing of large wind turbines. Project proposals will have to comply with the criteria indicated below.

2. Will select proposals for preliminary design studies (the Commission will contribute to the cost of each study with a lump sum).

3. Will select contractors for final design, construction, test operation and experimentation on a cost sharing basis, after completion of preliminary design studies.

In the evaluation of the projects, priority will be given to the targeted energy production cost (cost of kWh generated). Other non-technical factors will also be taken into account such as participants from at least two EC Member States, one or more major utilities as owner/operator, manufacturer with good record in wind energy, suitable expertise in estimating production costs and financial standing of proposer. Some indication of the firmness of funding support from sponsoring bodies would be helpful. The proposers shall provide the information requested in the Appendix.

Under stage (2) of the above procedure, the Commission hereby invites declarations of interest for design, development and testing of wind turbines, of at least 45m rotor diameter of equivalent swept area and an energy capture of at least 1.8GWh/y at the reference site (see Appendix). No restrictions are imposed on the configuration of rotational axis or number of blades.

Appendix

Information to be enclosed with the declaration of interest:

a) project objectives and plan for their achievement;

b) the indicative site(s) for the erection of the prototype and local wind characteristics;

c) names of possible participating organisations and Project Leader, including details of experience in wind energy;

d) expected performance and design information - including rotor diameter, estimates of annual energy output, theoretical performance curve, noise levels, weights of major components such as rotor, nacelle and tower;

e) information on transport;

f) preliminary estimate of the project costs for the prototype broken down into appropriate stages i.e. design, manufacture, installation, commissioning and testing;

g) forecast costs based on 50 machines/year over 10 years covering ex-factory costs of the turbine, foundation costs (assume no special provisions), installation and commissioning, operation and maintenance, electrical link from machine to site substation only.

Notes:

1. For estimation purposes a reference site at sea level will be assumed to have the following conditions:

- annual mean wind speed = 7.5m/s at 50m height
- wind speed variation with height based on 1/7 power law
- wind speed distribution assumed Weibull with k=2.0

The above is not intended to restrict the proposers' freedom of choice, however, proposers' estimates of performance at the reference site should be included.

2. All components must be of convenient size and suitable for transportation without special provisions being required. Procedure for erection must also be convenient.

3. 20-year life time and 100% availability (with no allowance for array efficiency) shall be assumed.

4. Energy costs referred to e) and f), excluding operation and maintenance costs, divided by the annual production at reference site (ECU/kWh/y).

5. Noise estimation according to IEA recommended practices, vol. 4. As mandatory target, the desired source power level (LWR) shall be less than $22 \log D + 72$ [dBA] where D = rotor diameter [m].

6. The power quality shall comply with IEA recommendations, vol. 7.

7. In general safety provisions shall comply with IEC/TC 88 (draft), safety philosophy of wind turbines generator systems.

2
Research and Development Projects within the JOULE-Programme

Bonus 1000kW

General Description

The general purpose of the project was to develop a commercially attractive large wind turbine. According to Bonus' experience the best economy could be achieved by stretching the company's earlier turbine designs, mainly the 450kW turbine. Therefore, the traditional elements of the „Danish Concept" are all found in the Bonus 750/1000kW. It has a three bladed rotor on the upwind side of the tower, constant speed stall control, a standard transmission system arrangement with a gearbox and an induction generator connected directly to the grid, active yawing and a fairly rigid tower.

The rotor diameter of the prototype is 50m and the nominal power rating is 1000kW. Commercially competitive gearbox and generator components are now available for 1MW power rating. It is likely that the 1MW rating will be the most cost efficient, therefore, after having tested the 750kW prototype, a 1,000kW version was built.

The **rotor blades** are made from glass fibre/polyester material and are manufactured by LM (LM 24). They use traditional profiles in the inner sections and more sophisticated FFA profiles in the outer part of the blade. A new lightning protection system was developed for the prototype rotor. It consists of an aluminium blade tip with an aluminium tip torpedo and an internal cable to the blade root. The three blades are mounted in four-point ball bearings.

The **rotor hub** is a ball-shaped cast design. All maintenance and bolt tightening is carried out inside the hub which has comfortable working space for two technicians.

Full span feathering is used as **aerodynamic brake**. This method turned out to be more cost-competitive than the rather expensive equipment for more traditional aerodynamic tip brakes. A failsafe system for the full span feathering system for aerodynamic braking was developed, based on three independent hydraulic units positioned in the rotor hub. Feathering is made in the deeper stall direction.

The **mechanical drive train** includes a main shaft in nodular cast iron. It is connected to the rotor hub with a flange and to the gearbox with a shrink disc. The gearbox is a hybrid planetary/helical gearbox with internal structure, tooth geometry and tooth grinding processes optimized in order to reduce the structural noise to the lowest possible level.

The **mechanical brake system** has four hydraulic disc brake calipers on the high speed shaft of the gearbox.

The **nacelle** is a self-supporting tubular steel structure. The rotor shaft is supported by one main upwind bearing, and the gearbox acts as the downwind bearing. This arrangement is similar to the design of the Bonus 300kW.

The **electric power conversion** system consists of an induction generator delivering 750kW at a level of 690V. The generator operates at a fixed speed of 1,500rpm. It is directly connected to the grid.

The **tower** of the prototype is a tapered tubular steel tower with a rotor hub height of 50m.

The Bonus 750kW development work started in mid 1993. The prototype was erected in September 1994. It was purchased by the Danish utility I/S Vestkraft and erected in Tjæreborg near Esbjerg in Denmark, close to the 2MW ELSAM turbine.

The Bonus turbine was erected in two days, 21 to 22 September 1994. On the first day the tower was erected and the rotor assembled on the ground. On the second day the nacelle was lifted onto the tower, and the rotor lifted up in one piece and bolted to the main shaft flange.

The first connection to the grid took place on 20 October 1994.

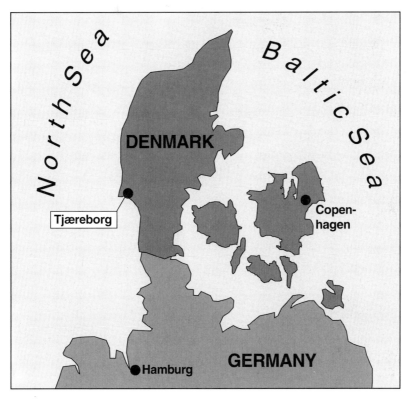

Bonus 750kW Site of the prototype at Tjæreborg, Denmark

Bonus 1000kW Prototype at Tjæreborg near Esbjerg, Denmark

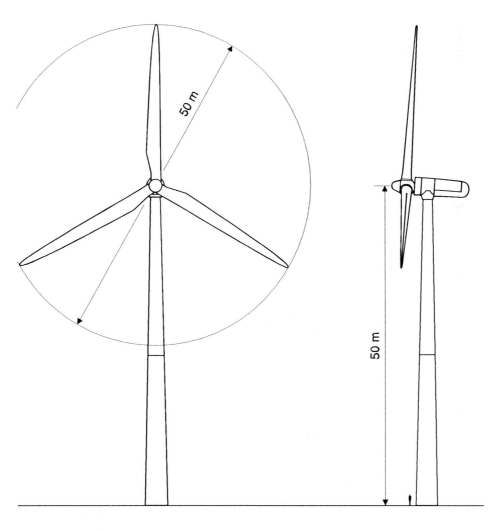

Bonus 750kW/1000kW Side and front view, 1:500

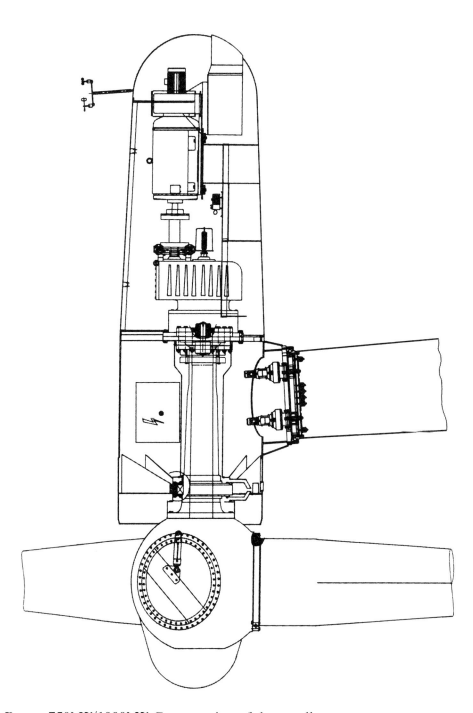

Bonus 750kW/1000kW Cross section of the nacelle

Main Data of the Bonus 750kW Prototype

Rotor
Number of blades 3
Orientation .... upwind
Diameter 50m
Swept area .... 1,963m^2
Hub height 50m
Tilt angle .... 4°
Rotational speed (fixed) 22rpm

Performance
Rated power output 750kW
Operational wind speeds:
        cut in 5m/s
        rated 15m/s
        cut out 25m/s
Max. tip speed 58m/s

Power Control
Type .... stall

Rotor brake
Mechanical brake:
Type .... disc
Position .... high speed shaft
Actuation .... hydraulic motors
Air brake:
Type .... full blade feathering
Actuation .... hydr. actuators in hub

Rotor Blades
Material .... glass fibre/polyester
Airfoil section NACA 634xx/FFA-W3

Hub
Type .... rigid
Material .... cast steel

Gearbox
Type .... 1x epicyclic; 2 x parallel
Ratio 1:69

Generator
Type .... induction
Rated power 750kW
Rated voltage 690V
Speed 1,500rpm

Yaw system
Actuation .... electric motors

Nacelle
Base frame ...... welded steel bedplate
Cover ..... steel frame and steel plates

Tower
Type .... tubular steel

Masses
Blade (each) 3.5t
Towerhead 59t
Tower 48t
Total 107t

Main Data of the Bonus 1000kW Prototype

Rotor
Number of blades 3
Orientation .... upwind
Diameter 54m
Swept area .... 2,290m^2
Hub height 50m
Tilt angle .... 4°
Rotational speed (variable) max.22rpm

Performance
Rated power output 1000kW
Operational wind speeds:
        cut in 5m/s
        rated 15m/s
        cut out 25m/s
Max. tip speed 62m/s

Power Control
Type .... stall

Rotor brake
Mechanical brake:
Type .... disc
Position .... high speed shaft
Actuation .... hydraulic motors
Air brake:
1. Type .... full blade feathering
Actuation .... el. actuators in hub
2. Type .... tip brake
Actuation .... hydr. or centrifugally

Rotor Blades
Material .... glass fibre/polyester
Airfoil section NACA 634xx/FFA-W3

Hub
Type .... rigid
Material .... cast steel

Gearbox
Type .... 1x epicyclic; 2 x parallel
Ratio 1:69

Generator
Type .... induction
Rated power 1000kW
Rated voltage 690V
Speed 1,500rpm

Yaw system
Actuation .... electric motors
Bearing .... sliding elements with

controlled frictionNacelle
Base frame .... steel bedplate
Cover ..... steel frame and steel plates

Tower
Type .... tubular steel

Masses
Blade (each) 4.2t
Towerhead 60t
Tower 48t
Total 108t

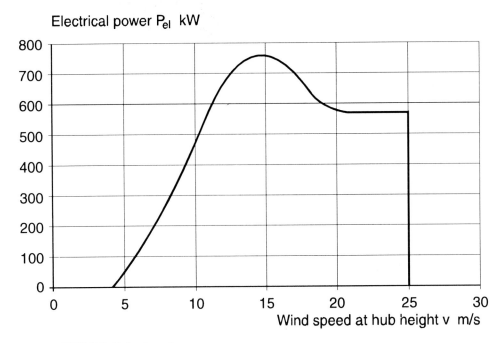

Bonus 750kW Calculated power curve

Bonus 750/1000kW Rotor hub

Bonus 750/1000kW Induction generator

Bonus 750kW Assembly of the rotor

Bonus 750kW Mounting of the nacelle

Bonus 750kW Mounting of the rotor

Bonus 750kW Prototype at Tjæreborg test site in Denmark

First Operational Results

The prototype of the 750kW turbine has been undergoing a comprehensive test programme since 1994. The test programme has three parts:

1. Running-in phase with limited automatic operation and stall regulation.
2. Measurement phase with unlimited automatic operation and stall regulation.
3. Variable speed phase with pitch regulation.

The measurement phase with stall regulation has been completed. At the time of writing, the variable speed phase has not yet been carried out.

During the stall regulation operation certain adjustments of the tuning of the structural dynamics had to be undertaken. The tuning was mainly adjusted by changes to various rubber elements in the structure. The structural loads have always been monitored and correspond to the loads calculated by Bonus. The power output, however, fluctuated considerably. 3P power fluctuations of several hundred kW were experienced. The problem was resolved by tuning and exchanging the generator.

Power Performance

The power curve has been measured. It corresponds to the calculated power curve.

Noise Emission

The noise emission has also been measured. The sound power level is 101dB at 8m/s 10m above ground. The value of 101dB (A) is satisfactory compared to the predicted value of 103dB (A) in the project proposal. The tonality of the turbine noise has also been found to be satisfactory with no risk of pure tone penalty under any of the noise regulations currently implemented in the EU.

Rotor Failure

In the evening of 17 October 1995 the 750kW prototype suffered a severe rotor failure due to an overspeed situation caused by simultaneous blocking of the three independent pitch systems. The rotor was destroyed and collision of the blade tips with the tower caused damage to the tower. The turbine has been dismanteled and the accident has been investigated by Bonus and a number of independent expert bodies. The investigation revealed problems in the blade pitch bearing arrangements of the 750kW prototype.

Bonus 1MW Prototype

Beginning of 1995 a Bonus 1000kW prototype was developed and as a consequence of the rotor failure of the 750kW prototype it was decided to erect the 1MW on the site, thus being the replacement in the WEGA-II project. The design of the 1MW turbine is based on the experiences gained from the former prototype. The turbine is stall controlled and with variable speed. The principle of the individual blade pitch system with full blade feathering into deeper stall is retained and additionally each blade is equipped with standard tip brakes, thus facilitating double failsafe blade pitching in stop sequences.

The turbine was erected in July 1996 and the commissioning phase started in August 1996.

Bonus 750kW Measured power curve
Diagram based on a table from Bonus dated 6 February 1996 (PLM bin, Rev.1.1)

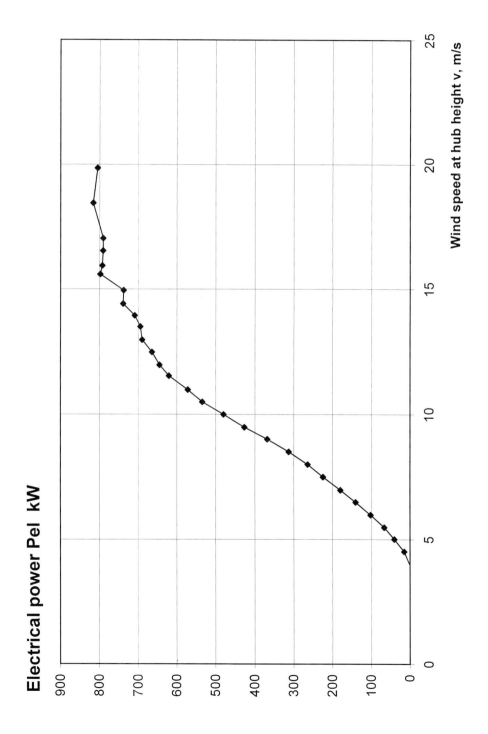

Enercon E-66

General Description

The Enercon E-66 features a gearless drive train concept as the only wind turbine in the megawatt class. Apart from this so far unique concept it is a three bladed horizontal axis pitch controlled wind turbine. The rotor diameter is 66m and the rated power 1,500kW. The design is based on the successful 500kW unit (E-40).

The **rotor blades** are of an advanced lightweight design (glass fibre epoxy material). The blade flange is connected to the rotor hub via a number of bolts laminated directly into the root structure of the blades. The rotor blades of the E-66 are equipped with an integrated lightning protection system to provide maximum safety for the blades as well as the bearings, the generator and the electronics. The rotor blades are mounted to the hub via ball bearing connections.

The rigid **hub** is connected to the stationary axle pin via tapered roller bearings. The rotor (rotating part) of the Enercon ring-generator is directly connected to the hub and does not need a bearing of its own.

The **blade pitch system** uses an electric system. The desired blade pitch angle is achieved by a pitch motor which changes the blade angle according to a pre-set value. Continuous measurement of the pitch angle ensures and controls the synchronous pitching of all blades. Each blade has its own actuator pitch control with an individual emergency unit that takes over the energy supply in case of a grid failure. If any disturbance of synchronism or other failure of the pitch system is detected the emergency braking process will be activated.

During normal operation – when the wind speed exceeds the rated wind speed – the rotational speed of the turbine is kept at a fixed value (rated speed) by changing the blade angles. The required adjustment of the blade pitch angles are determined by measurements of rotor speed and acceleration. In the partial load range the blade pitch angles are controlled according to the power output in order to achieve an optimized power output.

The turbine's **safety system** initiates quick pitching of the three rotor blades into feathered position. The energy supply of the pitch motors will be delivered by battery storages via the grid-supplied control system in case of a grid failure. The availability of these units is ensured by automatic recharging, load monitoring and load cycle testing. An emergency stop activates an additional hydraulic brake to stop the rotor completely.

The **electric system** consists of a multipole synchronous generator directly coupled to the rotor and an advanced frequency-inverter for variable speed operation. The use of the approved direct-driven ring-generator concept allows the omission of a gearbox. Its design is based on other Enercon wind turbine models. The essential design features are:

– a disc rotor with poleshoes mounted onto its outer diameter and
– a stator ring containing the generator windings.

The generator is corrosion protected with a coating of impregnating resin and an additional finishing varnish.

The **yaw control system** is always in operation for wind speeds above the lower limit (3m/s). The direction of the wind is continuously measured via a wind vane situated on top of the nacelle cover to ensure adjustment of the yaw operation with steady load bearing performance of the yaw drives avoiding momentum peaks during yawing.

The **tower** of the prototype is a steel tube with a rotor hub height of 68m. Special attention has been given to the issue of handling the components with fixtures and cranes readily available. Therefore, the generator stator and rotor have been winded and impregnated separately and have been subassembled before erection while the blades have been laminated in several parts and glued together after curing.

This concept of dividing the turbine into single components which are easy to handle regarding size and weight has also been applied for the erection. The tower is divided into four segments with a maximum weight of 30t for one segment and a maximum outer diameter of 4m (tower base). The erection of the turbine is carried out in several steps starting with the main carrier and the nacelle, then the generator stator which is followed by the generator rotor and finally the rotor hub with blade adaptors. Each blade is mounted separately by means of a winch located in the hub. After the first blade is fixed, the hub is rotated by means of the generator to enable the lifting of the second blade. The hub is rotated further and held in a fixed position by means of the generator while the third blade is mounted.

The site of the prototype is in Aurich (Ostfriesland) on the grounds of the Enercon factory. The prototype of the Enercon E-66 was erected at the beginning of December 1995 and connected to the grid for the first time in late December 1995.

Enercon E-66 Site at Aurich, Ostfriesland, Germany

Enercon E-66 - 1.5MW Prototype at Aurich, Germany

WEGA Large Wind Turbines

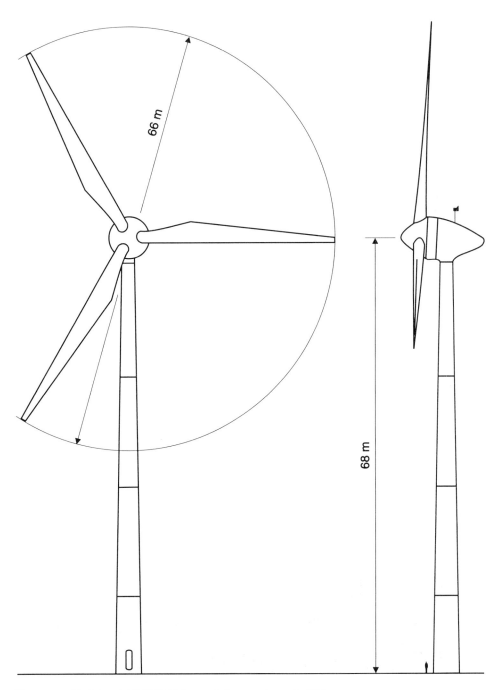

Enercon E-66 - 1.5MW Side and front view, 1:500

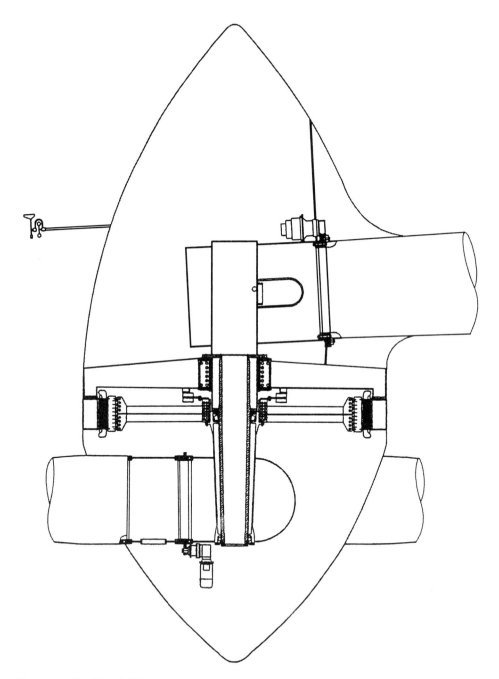

Enercon E-66 - 1.5MW Cross section of the nacelle

Main Data of the Enercon E-66 Prototype

Rotor
Number of blades 3
Orientation .... upwind
Diameter 66m
Swept area .... 3,421m^2
Hub height 68m
Tilt angle .... 3°
Rotational speed (variable) 10-20.3rpm

Performance
Rated power output 1,500kW
Operational wind speeds:
           cut in 2.5m/s
           rated 12.5m/s
           cut out 25m/s
Max. tip speed 70m/s

Power Control
Type .... full span blade pitch
Actuation .... three electric motors

Rotor brake
Mechanical brake:
Type .... hydraulic
Position .... low speed shaft
Air brake:
Type .... blade feathering

Rotor Blades
Material .... glass fibre/expoxy

Hub
Type .... rigid
Material .... cast steel

Generator
Type .... direct driven, speed variable,
.... low speed synchronous
Rated power 1,500kW
Rated voltage 460V
Speed range 10-20.3rpm

Yaw system
Actuation .... electrical

Nacelle
Base frame none (stationary axle pin)
Cover .... steel frame/
.... glass fibre polyester

Tower
Type .... tubular steel

Masses
Blade (each) 4t
Generator 48.6t
Towerhead 99.6t
Tower 120t
Total 219.6t

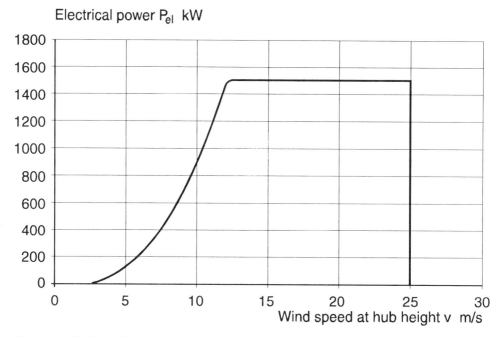

Enercon E-66 Calculated power curve

Enercon E-66 Manufacturing of the rotor blades

Enercon E-66 Manufacturing of the electric generator

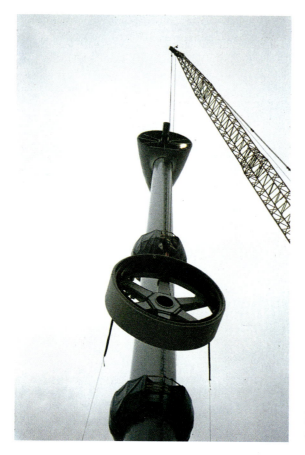

Enercon E-66 Mounting of the electric generator

Enercon E-66 Mounting of the nacelle

First operational results

In the third week of December 1995 the first rotation of the prototype took place. The turbine has been operating in automatic modes without staff supervision since the end of January 1996. Since 15 February 1996, the turbine has produced 176,988kWh in 469 hours of operation. Power performance and noise measurement started in February/March 1996.

Nordic 1000

General Description

The Nordic 1000 is a two bladed wind turbine which follows the example of the Nordic 400/500 with a lighter and less complex design than the traditional wind turbines.

The rotor diameter of the Nordic 1000 is 53m and the rated power output 1,000kW. One of the features leading to a simple design is the stall controlled rotor in order to limit the power level at high wind speeds. Other important simplifications are the design of the yaw system which does not need the expensive yaw brakes normally built into large wind turbines and the integration of the main shaft bearings in the gearbox.

The **rotor blades** are made of glass fibre reinforced polyester material. They have been developed and manufactured by LM, Denmark.

The teetering movements of the **hub** are counteracted by rubber bumpers which restrict the movements to a few degrees. This is enough to reduce the loads to a level which makes the design as resistant to fatigue as to extreme loads. The hub is compact with a comparatively low weight and a short overhang. It is cast in nodular iron and designed for infinite life.

The **mechanical drive train** is compactly designed. The planetary gear-box contains the main shaft bearings which makes the wind turbine very compact. The lubrication is of a splash type, with cooling provided by a separate oil cooler. The cooler fan also operates in case the temperature inside the nacelle becomes too high.

The secondary shaft carries the brake disc and two flexible couplings, one of which has a torque limiting capability. This safe-set coupling is mounted between the generator and the brake disc, i.e. the mechanical brake is also serviceable when the coupling is released. The two brake calipers utilise brake pads of a non-asbestos type. They are engaged by the force of springs in case the hydraulic pressure is released (passive system). The control system indicates excessive wear of brake pads and the complete non-engagement of the brake pads. The machinery bed is welded from cylindrically shaped sections which makes it compact and easy to manufacture. The gearbox is flanged to the front end of the bed, the generator to the shaft section; downwards the bed is connected to the yaw bearing.

The **electrical system** in the prototype unit consists of an induction generator and an AC-DC-AC system for variable speed. In the serial version, the generator will be a two-speed induction machine directly connected to the grid.

The **nacelle** provides a shelter for maintenance personnel and equipment with full standing height. The top and the bottom half made of glass fibre reinforced polyester are bolted together. The walls and roof are covered with sound absorbing material. The floor is covered with an anti-slip rubber mat. Intakes for ventilation in front of the nacelle are designed so as to avoid noise diffusion. The ventilated air is exhausted through the oil cooler and the generator. The latter one is equipped with a noise suppressor. The nacelle is fixed via vibration dampers.

There is a hoist on top of the machinery bed which can be used to take equipment etc. through a hatch in the aft end of the nacelle or through the tower. It can also be used to lower a blade tip to the ground. The hatch serves also as an emergency exit in case the evacuation equipment is being used. The machinery bed and the tower top are connected by a yaw bearing which makes it possible to position the turbine in the wind.

The **yaw drive** consists of two geared hydraulic motors which act on the inner toothed ring of the yaw bearing. The hydraulic pump unit also supplies the secondary shaft mechanical brake. The system is provided with temperature sensors, heater and cooler. During all operations of the yaw system, the turbine can perform small, damped rocking movements. There is no yaw brake.

The **controller** at the bottom of the tower as well as in the nacelle is based on standard type microprocessor control. The data is transferred by means of a serial interface with an optical link. The system can be entered into by modem and PC.

The **conical cylindrical tower** is welded in steel in one piece. The concrete foundation is secured to the limestone rock by 11m long pre-stressed bars.

The basic design of the Nordic 1000 is certified by DNV.

The site of the prototype is on the wind turbine test field on the island of Gotland, Sweden. The turbine was erected in April 1995. The first connection to the grid took place in June 1995.

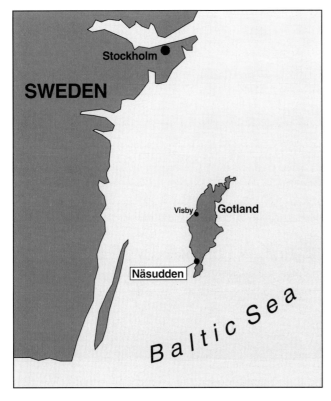

Nordic 1000 Site of the prototype on the island of Gotland near Näsudden, Sweden

Nordic 1000 - 1MW Prototype at Näsudden on the island of Gotland, Sweden

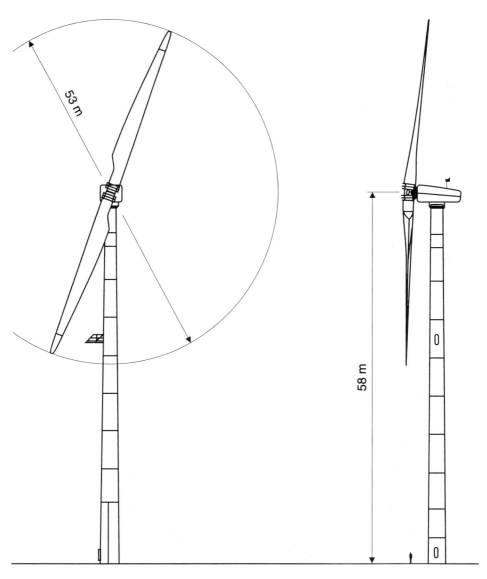

Nordic 1000 Side and front view, 1:500

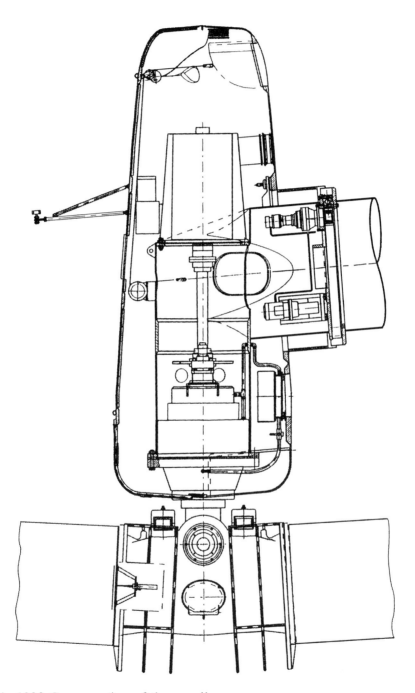

Nordic 1000 Cross section of the nacelle

Main Data of the Nordic 1000 Prototype

Rotor
Number of blades 2
Orientation .... upwind
Diameter 53m
Swept area 2,206m²
Hub height 58m
Tilt angle .... 4°
Rotational speed (variable) 15-25rpm
Cone angle .... 3°

Performance
Rated power output 1,000kW
Operational wind speeds:
                 cut in 4m/s
                 rated 16m/s
                 cut out 21m/s
Survival wind speed 50m/s
Max. tip speed 69.4m/s

Power Control
Type .... stall

Rotor brake
Mechanical brake:
Type .... disc
Position .... high speed shaft
Actuation ... spring released/hydraulic
Air brake:
Type .... tip brakes
Actuation .... centrifugally released/
.... hydraulic

Rotor Blades
Material .... glass fibre/polyester
Airfoil section
.... LM2/NACA 63.4xx/FFA-W3

Hub
Type .... teeter
Material .... nodular iron

Gearbox
Type .... 2 x epicyclic
Ratio 1:51.7

Generator
Type .... induction
Rated power 1,000kW
Rated voltage 690V
Speed (nominal) 1,307rpm

Yaw system
Actuation .... hydraulic

Nacelle
Base frame ... load carrying steel tube
Cover .... glass fibre/polyester

Tower
Type .... tubular steel
Diameter (Top/Base) 1.88/2.62m

Masses
Blade (each) 4.25t
Towerhead 42t
Tower 50t
Total 92.97t

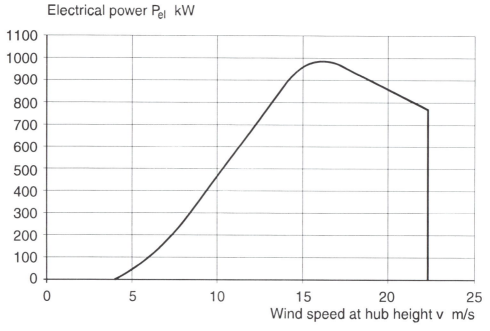

Nordic 1000 - 1MW Calculated power curve

Nordic 1000 Assembly of the nacelle

Nordic 1000 Transport of the tower

Nordic 1000 Nacelle and rotor

First operational results

During the commissioning, the Nordic 1000 has been in automatic operation most of the time, surveyed by Vattenfall personnel at the site and by remote control. Restrictions on power level have been gradually removed. So far (May 1996) it has logged 4,200 operational hours and produced 715MWh. It has been running at all power levels including rated power 1MW. A preliminary power curve has been composed from running at different rpm.

The commissioning has passed without any major problems. Most concern has been devoted to the electrical system, consisting of standard AC-DC-AC self commutated converters based on GTO thyristors, PWM technology and vector control of the induction generator. Initially there were problems with running the converter units in parallel, which meant that the power level had to be restricted to about 400kW. Finally these problems were solved by trimming the parameters of the converter system.

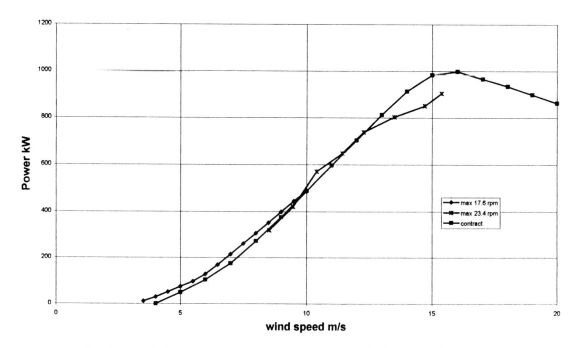

Nordic 1000 Preliminary power curve composed from different runs at maximum rpm of 17.6 and 23.4

Vestas V 63 - 1.5 MW

General Description

The Vestas 1.5MW wind turbine is a three bladed, pitch controlled horizontal axis wind turbine with an upwind rotor. The nominal rotor diameter is 63m and the rated electric power output 1,500kW. The rotor hub height of the prototype is 60m. The design is based on the experience gained from the V39 - 500kW and V42 - 600kW machines in mass production.

The **rotor blades** are of an advanced lightweight design and construction. The basic structure is a laminated glass fibre/epoxy material. The connection to the rotor hub is made by a lightweight aluminum flange bound to the glass fibre compound material.

The **rotor hub** is cast steel. For pitching the rotor blades the hub contains a set of ball bearings and the blade pitch mechanism.

The **blade pitch system** features the individual pitching of each blade. This provides a high degree of safety in any emergency situation. Moreover, individual pitching gives the opportunity of further optimization in future in case of individual controlling of each blade pitch angle. The Vestas OptiTip feature is standard for all Vestas 1.5MW turbines. This feature provides the optimal tip angle at any time for both power performance and noise emission. It enables Vestas to supply special versions with lower noise emission. The three actuators in the pitch system are powered by a hydraulic pump placed in the nacelle.

The **rotor shaft** and **bearing system** comprise a bedplate mounted arrangement with two separate roller bearings.

The **gearbox** is a combined planetary and parallel stage design with a gear ratio of 72. It is supported by the rotor shaft and torque suspended by the bedplate.

The **electric generator** is an induction generator with an „on side" speed variation capability through the so-called OptiSlip system. By this system the rotor of the generator can „slip" up to 10% of the nominal speed at rated power only. The slip power is dissipated in a system of resistances connected to the rotor. The system is controlled in such a way that only power peaks due to wind gusts trigger the slipping of the rotor. The OptiSlip generator's feature is to keep the power nominal during high wind speeds independent of air temperature and density. The use of the OptiSlip concept secures smooth power output and reduces the loads significantly.

The **nacelle** of the wind turbine consists of a compact load carrying welded steel bedplate. The cover is made of semi-forced glass fibre material.

The **tower** is a tapered steel tube with a standard height of 60m. The concept of the Vestas 1.5MW turbine includes two different sizes of rotor diameter: 57m and 63m respectively. Each version is designed to be erected on sites with a rather high or low wind regime. The V63 - 1.5MW wind turbine is manufactured in modular sections enabling the transport of e.g. the nacelle in standard containers. Thus no special equipment is necessary when transporting the nacelle from the factory to the site.

During the erection the maximum lift to hub height is about 25 tons because of the modular construction of the nacelle. The weight of the tower section is adjusted, so that a crane with a capacity of 25 tons is capable of lifting the tower, too. Thus, the crane capacity must not be unrealistically high.

The first prototype has been erected on the wind turbine test site near Tjæreborg, Denmark. The first connection to the grid was on February 9, 1996.

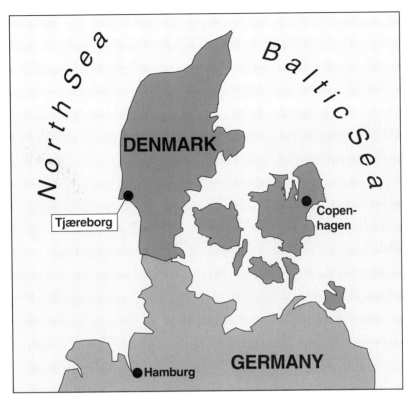

Vestas V63 - 1.5MW Site of the prototype at Tjæreborg, Denmark

Vestas V63 - 1.5MW Prototype at Tjæreborg, Denmark

WEGA Large Wind Turbines

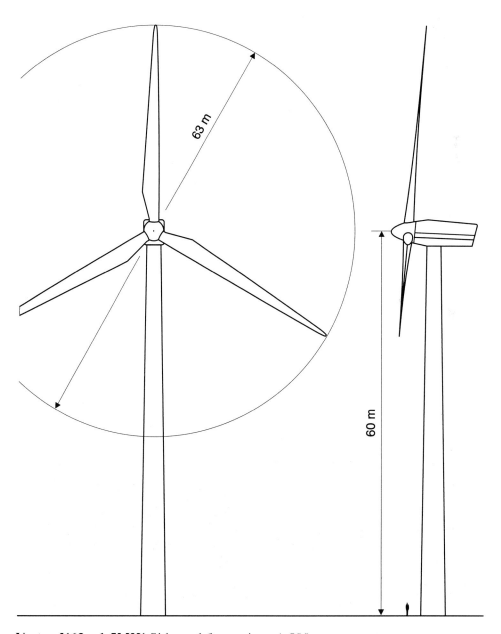

Vestas V63 - 1.5MW Side and front view, 1:500

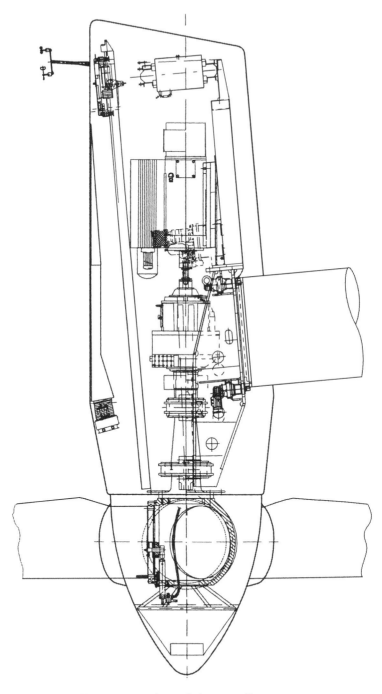

Vestas V63 - 1.5MW Cross section of the nacelle

Main Data of the Vestas V63 - 1.5MW Prototype

Rotor
Number of blades 3
Orientation .... upwind
Diameter 63m
Swept area .... 3,117m^2
Hub height 60m
Tilt angle .... 5°
Rotational sp. (variable) 20.8-22.9rpm

Performance
Rated power output 1,500kW
Operational wind speeds:
                 cut in 4.5m/s
                 rated 16m/s
                 cut out 25m/s
Max. tip speed 75.5m/s

Power Control
Type .... blade pitch
Actuation .... hydraulic

Rotor brake
Mechanical brake:
Type .... disc
Position .... high speed shaft
Actuation .... hydraulic
Air brake:
Type .... blade feathering
Actuation ..... 3 hydr. cylinders in hub

Rotor Blades
Material .... glass fibre/epoxy
Airfoil sect. .NACA 63.xxx/FFA-W3

Hub
Type .... rigid
Material .... cast steel

Gearbox
Type .... 1x epicyclic; 2 x parallel
Ratio 1:72.15

Generator
Type .... induction
Rated power 1.500kW
Rated voltage 690V
Speed range 1,500-1,650rpm

Yaw system
Actuation .... electric

Nacelle
Base frame ...... welded steel bedplate
Cover .... glass fibre polyester

Tower
Type .... tubular steel
Diameter (Top/Base) 2.0/4.0m

Masses
Blade (each) 3.5t
Rotor (incl. hub) 21t
Nacelle 52t
Towerhead 73t
Tower 80t
Total 153t

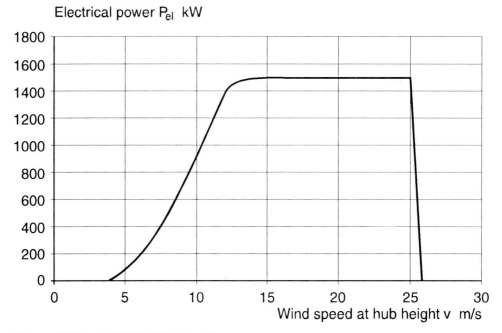

Vestas V63 - 1.5MW Calculated power curve

Vestas V63 Rotor hub with individual blade pitch system

Vestas V63 Rear of the nacelle with transformer

Vestas V63 Erection of the prototype at Tjæreborg, Denmark

Mounting of nacelle   Main shaft/gearbox   Generator

WEG MS4 - 600kW

General Description

The MS4 - 600kW is a three bladed, stall-controlled wind turbine with innovative features. The basic concept is derived from the two bladed wind turbine developed and manufactured by Carter in the USA. The rotor diameter of the MS4 is 41m and the rated power output 600kW.

The main feature is that it is a downwind machine with highly flexible blades. The nacelle is elongated so that the gearbox and generator balance the rotor, and the nacelle mounted to the tower is soft in yaw and pitch, both being heavily damped. The effect of this compliant arrangement is that the advantages usually associated with free teeter are available from a three bladed rotor, with considerable reductions in loads in all the structural elements.

The **rotor blades** run 7 - 8 tower diameters downwind of the tower where the flow has re-combined and the blade thump normally associated with downwind operation does not exist. The flexibility of the blades reduces operating mean loads to 25% compared with stiff blades and storm loadings to 50%. The rotor is stall controlled with a narrow tip chord and 64m/s tip speed, giving low aerodynamic noise. Rotor braking is being made by whole-blade pitching towards stall, under the action of individual hub-mounted failsafe actuators.

The **mechanical drive train** i.e. the main shaft and gearbox assembly has a 3-point mounting to the nacelle, the two rear mounts being soft rubber. Again, this is a current desing trend for low noise in large turbines in order to provide good mechanical isolation of the gearbox.

The **nacelle tube** is off-set from the tower and slung below the yaw bearing: this enables the nacelle and rotor to be raised and lowered from the tower top for installation and maintenance. The winch used for this purpose also serves to erect the tower initially.

Another main feature is the damped **yawing** and **nodding** so that the rotor can follow rapid wind direction changes smoothly without neither: a) operating off-wind, as in the case of a conventional machine yawing too slowly, nor b) incurring gyroscopic loads, as with a conventional machine yawing too rapidly. By nodding the rotor, it can also follow changes in wind shear and pitch. The concept almost halves the weight of the nacelle and the rotor (slightly more than 20t) as compared with equivalent conventional designs. The use of standard components, e.g. gearbox and generator, means that weight reduction is reflected in lower costs.

The **electrical power system** is an induction generator with a 1% slip at rated power and is directly connected to the grid.

The **tower** of the MS-4 is a very slender steel tube tower. It also serves as a crane for hoisting the rotor and the nacelle by means of a simple winch.

The MS-4 is particularly suited for "difficult" wind farm sites with complex topography where high turbulence and gusts up to 70m/s are combined with poor access. Noise levels are state-of-the-art, and low weight ensures that the installed costs are low. The whole turbine is of slender proportions and therefore attractive for planning authorities. The nacelle and tower can be fitted into standard containers. The machine has its own winching system so that it can be installed with only a small off-loading crane. There is full man access to the nacelle, and the winching system can be used for the replacement of major components. The machine is certified according to the design standards of Germanischer Lloyd.

At present, the MS4 - 600kW has not yet been installed. As of February 1996, component tests are well under way. The operation of a 1/20 scale model has confirmed computer predictions of the stability of the machine under a variety of running conditions. The plan is to install the prototype at a rough site in Wales this year and to operate and monitor it during the year 1997.

WEG MS4 - 600kW Artist's view of the prototype

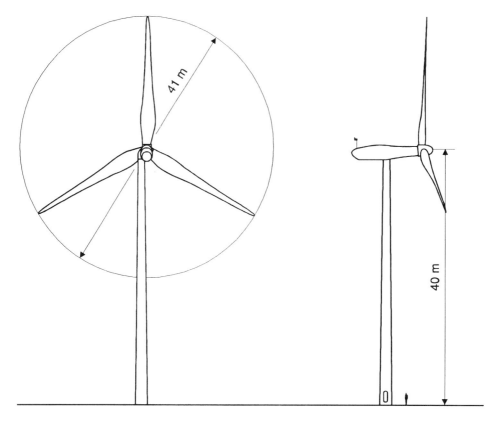

WEG MS4 - 600kW Side and front view, 1:500

WEGA Large Wind Turbines

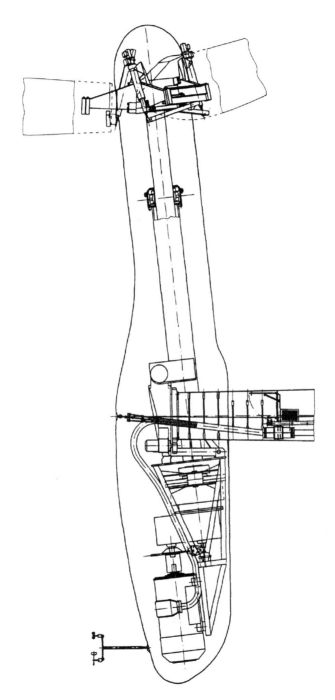

WEG MS4 - 600kW Cross section of the nacelle

Main Data of the WEG MS4 - 600kW Prototype

Rotor
Number of blades 3
Orientation .... downwind
Diameter 41m
Swept area 1,320m²
Hub height 40m
Tilt angle .... 6°
Rotational speed (fixed) 29.6rpm

Performance
Rated power output 600kW
Operational wind speeds:
        cut in 5m/s
        rated 14m/s
        cut out 25m/s
Max. tip speed 64m/s

Power Control
Type .... stall

Rotor brake
Mechanical brake:
Type .... disc
Position .... high speed shaft
Actuation .... spring applied
Air brake:
Type ...full span negative blade pitch
Actuation ..... 3 hydr. actuators in hub

Rotor Blades
Material ..glass fibre polyester/epoxy
Airfoil section .... NACA 63xxxx

Hub
Type .... rigid
Material .... steel

Gearbox
Type .... 1x epicyclic; 2 x parallel
Ratio 1:57.2

Generator
Type .... induction
Rated power 600kW
Rated voltage 690V
Nominal speed 1,500rpm

Yaw system
Actuation .... hydraulic

Nacelle
Base frame ...load carrying steel tube
Cover .... glass fibre/polyester

Tower
Type .... steel tube

Masses
Blade (each) 1.68t
Rotor 5t
Towerhead 21.6t
Tower 23.7t
Total 45.3t

WEGA Large Wind Turbines

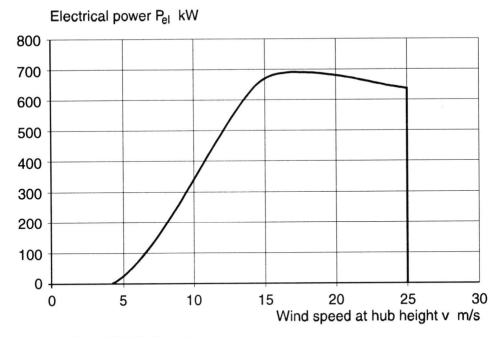

WEG MS4 - 600kW Calculated power curve

3
Demonstration Projects Supported by THERMIE Evaluated Within the WEGA-II Programme

AEOLUS II, NÄSUDDEN II

General Description

The **AEOLUS II** and **NÄSUDDEN II** wind turbines are following the multi-megawatt development in Sweden and Germany. The turbines have been built with the know-how gained during the NÄSUDDEN 2MW project from 1978 to 1987.

From 1987 - 1988 Kvaerner Turbin AB carried out a concept study in order to increase the availability of the turbines and to decrease at the same time the O&M costs. Cost reduction in general and system simplification were of course other important design parameters.

Swedish and German manufacturers, the government and public utilities formed a cooperation for the further development of large wind turbines. Consequently, Preußen Elektra and Vattenfall ordered one turbine each for Germany and Sweden. In this project MBB was responsible for the blade design and manufacturing, for the aerodynamic and dynamic design and for the delivery of the German turbine. Kvaerner Turbin AB was responsible for the design of the structural and machinery parts and the delivery of the Swedish turbine. The order was placed at the beginning of 1989 and the turbines were connected to the grid in 1993. NÄSUDDEN II has been erected on Gotland in the Baltic Sea, AEOLUS II in Wilhelmshaven at the German North Coast.

AEOLUS II and NÄSUDDEN II feature the same overall concept. They are both blade pitch controlled horizontal axis wind turbines. The rotor diameter is 80m and the rated power output 3,000kW. However, there are two important differences: AEOLUS II has a newly designed tower whereas the NÄSUDDEN turbine uses the old concrete tower from NÄSUDDEN I.

Moreover, AEOLUS II has a variable speed whereas NÄSUDDEN features a two-speed concept.

The **rotor blades** are made of load carrying parts of carbon fibre/epoxy as well as leading and trailing parts of glass fibre/epoxy. The shear web and stringers are made of PVC foam which is used inside the spar to carry shear loads and prevent buckling. A copper mesh covering all carbon fibre serves as lightning protection.

The **hub** consists of two welded steel tubes in T-form. The connection to the turbine shaft is a hydraulically assembled OK-coupling (SKF). The hub contains parts of the blade pitch system. It has also an external balcony for maintenance of the blade bearings.

A full span **pitch system** performs above rated power, 3MW. Blade pitching is carried out by a hydraulic servocylinder behind the gearbox via a rod through the gearbox and turbine shaft into the hub. A yoke then transmits the force to each blade. The blade bearings are of a ball bearing type. The blade pitch system is also the main braking system. This brake is activated through a separate hardwire system in case the turbine is overspeeding.

The **mechanical drive train** is placed on a bedplate of welded steel. The gearbox is part of the load carrying structure as the turbine shaft is integrated into the gearbox housing which is bolted to the bedplate. Thus, the turbine shaft bearings are part of the gearbox lubrication and the length of the turbine shaft is minimised. The gearbox has two planetary gears and a bevelled gear as endstage. It has a pressurized lubrication system with filtering and cooling.

The **electric power conversion system** is different in the two wind turbines. AEOLUS II features a speed variable AC-DC-AC system on the basis of a synchronous generator. The speed range is approximately -50% to +15% from the rated speed. NÄSUDDEN II has a two-speed system on the basis of an induction generator. The generator has double windings with six and four poles for two constant rotor speeds. This was chosen mainly for noise reasons as the machine can be operated at a lower tip speed at low wind speed. Between the gearbox and the generator there is a break coupling, protecting the gearbox from a possible short circuit of the generator. The generator has a slip ratio of 3% to provide better dynamic performance of the drive train. The solution from NÄSUDDEN I with a stationary generator in the tower top was maintained for NÄSUDDEN II and AEOLUS II. Thus, a transmission system for the power cables from a rotating nacelle is avoided.

The **nacelle cover** is a sandwich design with an outer skin of aluminium. Important design work has been carried out to decrease noise emission. The nacelle cover and the cylindrical steel tower top are noise insulated. The nacelle cover is also isolated from the tower by means of rubber dampers. Wind measurement equipment is placed on top of the cover.

The turbine is kept in the direction by mechanical yaw brakes. **Yaw** is performed by means of two hydraulic motors placed in the nacelle acting on a gear ring fixed in the tower. The yaw bearing is also of a slewing ring type.

The **tower** of NÄSUDDEN II is the same as the tower used for the first NÄSUDDEN 2MW turbine. The tower is made of concrete with prestressed wires. It stands on a gravity foundation as the limestone at this site is of poor quality. On the outside of the tower, rails are fixed in order to lift the nacelle to the tower top. Inside the tower there is a lift for transport. The dynamic concept is based on the NÄSUDDEN I. It is a dynamically stiff design with bending frequencies well above the critical 2P-frequency.

For AEOLUS II a new concrete tower was built on the site. Its dynamic concept is „soft" i.e. the first bending eigenfrequency of the tower is around 1.5P.

AEOLUS II has been installed in the Jade Wind Energy Park near Wilhelmshaven at the northern coast of Germany. The turbine was commissioned in December 1993.

The site of NÄSUDDEN II is the island of Gotland (Sweden), the same location where the former 2 MW turbine has been tested. The wind turbine was connected to the grid in March 1993.

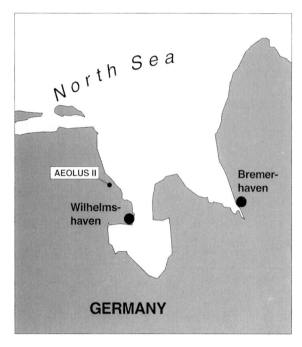

AEOLUS II 3MW Prototype in the Jade Wind Park, Wilhelmshaven, Germany

NÄSUDDEN II 3MW Site of the prototype near Näsudden, Gotland

AEOLUS II Prototype at the Jade Wind Park near Wilhelmshaven, Germany

NÄSUDDEN II Prototype on Gotland, Sweden

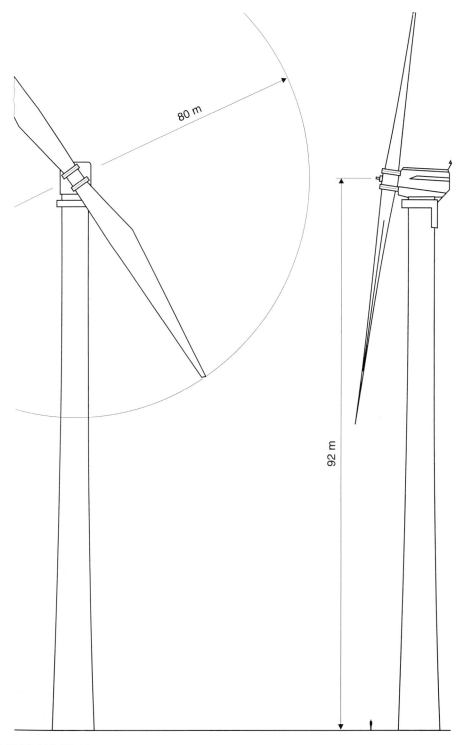

AEOLUS II Side and front view, 1:500

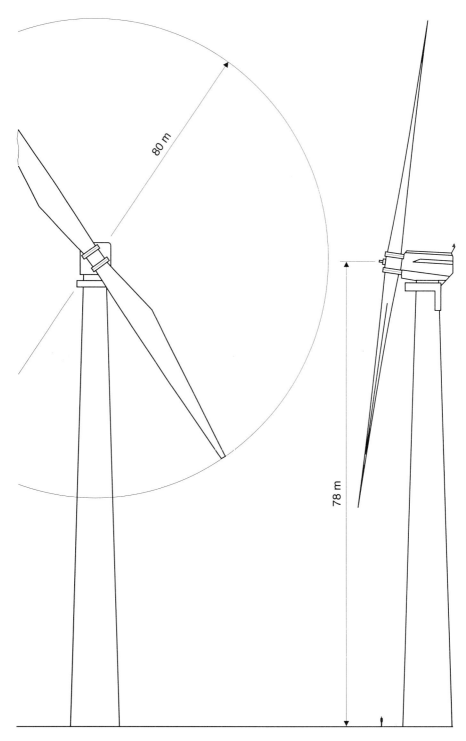

NÄSUDDEN II Side and front view, 1:500

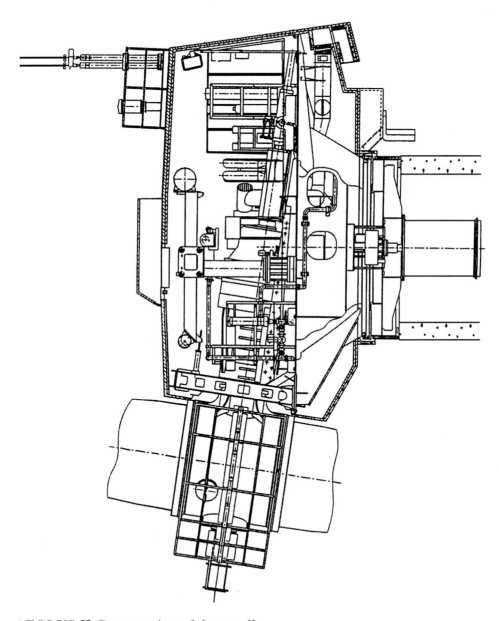

AEOLUS II Cross section of the nacelle

Main Data of the AEOLUS II Prototype

Rotor
Number of blades 2
Orientation ....upwind
Diameter 80.5m
Swept area............................5,090m^2
Hub height 92m
Tilt angle ....8°
Rotational speed (variable) 10.5-24rpm

Performance
Rated power output 3,000kW
Operational wind speeds:
         cut in.................6m/s
         rated...........14.15m/s
         cut out.............25m/s
Survival wind speed 75m/s
Max. tip speed ....100 m/s

Power Control
Type ....full span blade pitch
Actuation ....hydraulic

Rotor brake
Mechanical brake:
Type...disc
Position.....................low speed shaft
Actuation ....hydraulic/
.....................................spring applied
Air brake:
Type ....full span blade pitch
Actuation ....hydraulic in nacelle

Rotor Blades
Material ......glass fibre/epoxy/carbon
Airfoil section.....................FW84-W

Hub
Type...rigid
Material ....steel

Gearbox
Type.....2xepicyclic/1xbevelled stage
Ratio 1:74

Generator
Type ....synchronous
Rated power........................3,000kW
Rated voltage 6,000V
Speed range ....780 - 1,550rpm

Yaw system
Actuation ....hydraulic

Nacelle
Base frame ......welded steel bedplate
Cover ....aluminum

Tower
Type .....tubular/pre-stressed concrete

Masses
Blade (each)................................9.5t
Rotor (incl. hub)..........................41t
Nacelle..121t
Towerhead 162t
Tower......................................1,410t
Total..1,572t

Main Data of the NÄSUDDEN II Prototype

Rotor
Number of blades 2
Orientation ....upwind
Diameter 80.5m
Swept area ....5,090m^2
Hub height 78m
Tilt angle ....8°
Rotational speeds (fixed)...14/21rpm

Performance
Rated power output 3,000kW
Operational wind speeds:
   cut in................6m/s
   rated...............14m/s
   cut out............25m/s
Survival wind speed 75m/s
Max. tip speed 88.5m/s

Power Control
Type ....full span blade pitch
Actuation ....hydraulic

Rotor brake
Mechanical brake:
Type...disc
Position....................low speed shaft
Actuation ....hydraulic/
......................................spring applied
Air brake:
Type ....full span blade pitch
Actuation ....hydraulic in nacelle

Rotor Blades
Material ......glass fibre/epoxy/carbon
Airfoil section.....................FW84-W

Hub
Type...rigid
Material ....steel

Gearbox
Type.....2xepicyclic/1xbevelled stage
Ratio 1:73.47

Generator
Type ....induction
Rated power................750/3,000kW
Rated voltage.......................6,000V
Nominal speed 1,000/1,500rpm

Yaw system
Actuation ....hydraulic

Nacelle
Base frame ......welded steel bedplate
Cover ....aluminum

Tower
Type ......tubular/prestressed concrete

Masses
Blade (each)...............................9.5t
Rotor (incl. hub).........................41t
Nacelle......................................121t
Towerhead 162t
Tower....................................1,510t
Total.....................................1,672t

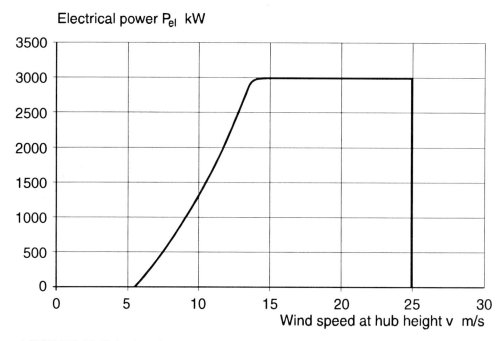

AEOLUS II Calculated power curve

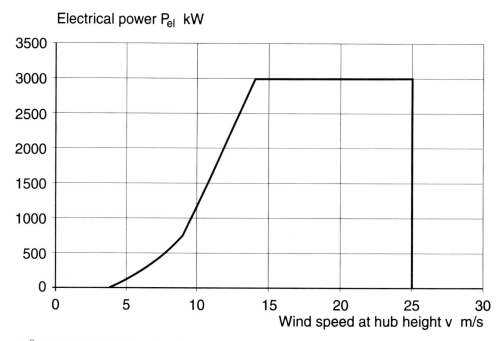

NÄSUDDEN II Calculated power curve

AEOLUS II Manufacturing of a rotor blade

AEOLUS II Rotor hub with nacelle

AEOLUS II Mounting of the nacelle

AEOLUS II Mounting of a rotor blade and the nacelle

First Operational Results

Both wind turbines have been tested for more than two years now. Thus the operational record covers a longer period.

NÄSUDDEN II

The NÄSUDDEN II was connected to the grid in March 1993 for the first time. Since then it has produced more than 15,000MWh with an average availability of about 85%. The availability reached 97 - 99% during the following months. One of the initial goals, to increase the availability, has been reached. The technical problems so far have been mainly related to the mechanical brakes. Some unexpected dynamic behaviour has also been encountered, mainly due to the big tower mass.

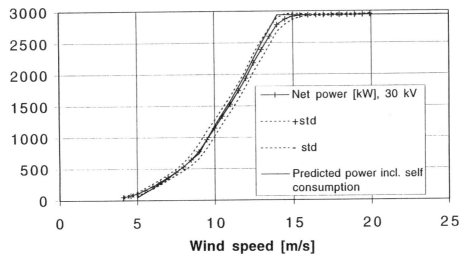

Näsudden II Diagram based on table FFA/RNG-960219

AEOLUS II

During the first year of operation the staff of Preussen Elektra AG was trained in repair and maintenance. Measurements have been carried out and modifications have been tested, too. Starting from December 1993 the net production reached 5,300MWh in one year. Excluding the time of testing and measurement the machine has reached the foreseen power output and an availability of 94%.

The predicted energy production at the site is about 7,000MWh. Basically the measured power curve is close to the calculated power curve. Noise emission measurements have also been performed.

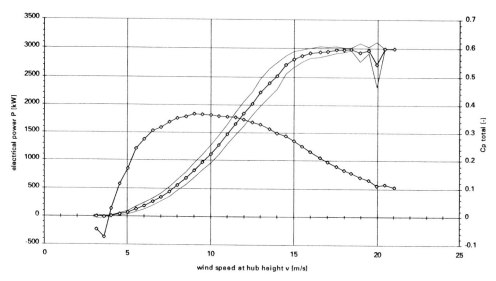

AEOLUS II Measured power curve and power coefficient (DEWI, 1995)

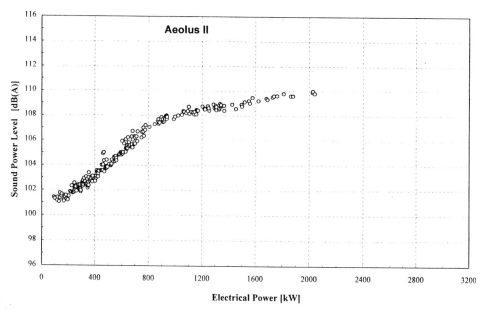

AEOLUS II Noise emission (DEWI, 1995)

West GAMMA 60

General Description

The GAMMA 60 is a two bladed wind turbine with an upwind rotor and a teetered hub. The wind turbine has been designed by AERITALIA/WEST under joint contract with ENEA and ENEL. The design characteristics of the GAMMA 60 were customised to meet the requirements of the Italian electric utilities. The rotor diameter of the GAMMA 60 is 60m and the rated power output of the prototype is 1,500kW.

The advanced concept of the GAMMA 60 includes two main innovative elements; the first one is the yaw drive system utilised for power regulation. By this procedure, the blade pitch system can be eliminated, thus reducing the costs of many components in the machine. The second innovative element is the operation in a broad range of variable speed modes in order to increase energy capture and fatigue life.

The **rotor blades** are made of glass fibre reinforced polyester material. Each blade is 29m long. The blades are manufactured in filament wound technology. The whole blades have been machined by means of a large winding machine at Hamilton Standard in the USA.

Elastomeric bearings and bumpers in the **steel hub** enable the rotor to teeter, thus reducing the loads which makes the concept more resistant to fatigue and extreme loads.

The **mechanical drive train** consists of a main shaft manufactured of a single block of forged steel supported by roller bearings in a very strong housing. It is hollow so that signal wires can pass through. The rotor torque is transmitted to the electric generator through the main shaft, the gearbox and the high speed shaft. The gearbox is a two-stage parallel design with a gear ratio of 32.57. It is supported by the rotor shaft and torque suspended by the bedplate.

The **yaw drive system** includes two hydraulic motors and the roller bearing support of the nacelle. Controlled yaw is used instead of pitch control for power control. This design is used for less sophisticated machines for the purpose of both manufacture and maintenance.

The **electrical system** includes a synchronous generator operating at variable frequency and voltage. A static voltage and frequency converter is used for the connection to the grid. The command and monitoring system ensures that control is maintained both in normal and in emergency operating modes. A DC link is used between the generator/rectifier and the inverter, because this type of frequency converter system provides the necessary speed range, excellent torque control and electrical damping of structural modes. It also avoids the stability problems which would be associated with some of the split-path frequency conversion systems proposed for wind turbines.

The **tower** is a free-standing steel tube design. Inside the tower, power and signal wires are located. A lift provides access to the nacelle.

The first prototype was erected at Alta Nurra on Sardinia and was connected to the grid in May 1992.

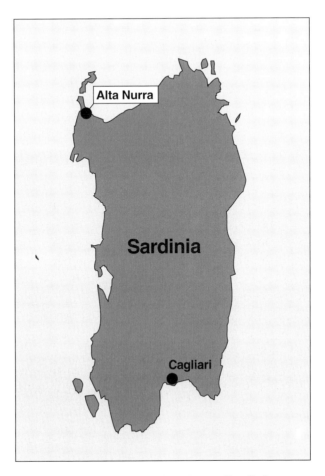

GAMMA 60 Site of the prototype at Alta Nurra, Sardinia

GAMMA 60 Prototype at Alta Nurra, Sardinia

WEGA Large Wind Turbines

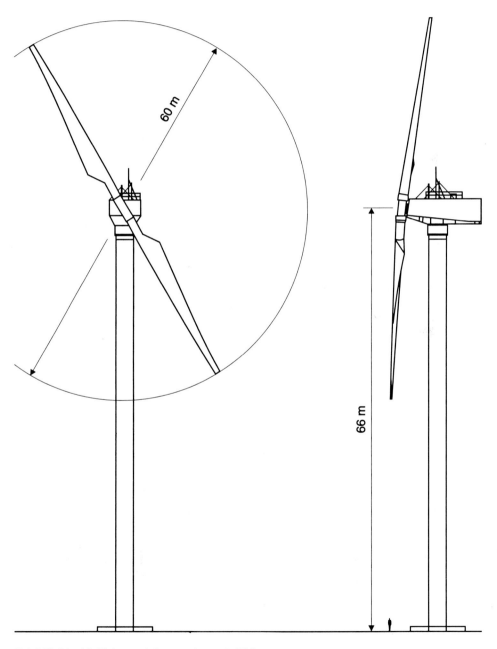

GAMMA 60 Side and front view, 1:500

88 WEGA Large Wind Turbines

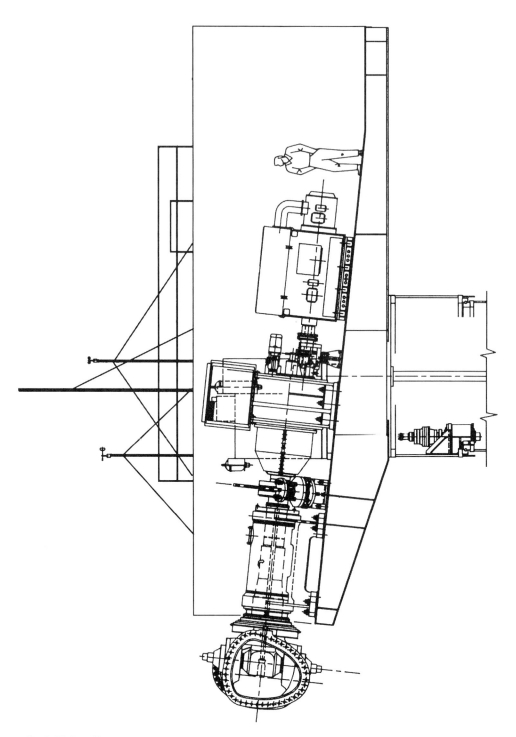

GAMMA 60 Cross section of the nacelle

Main Data of the GAMMA 60 Prototype

Rotor
Number of blades 2
Orientation .... upwind
Diameter 60m
Swept area .... 2,827m^2
Hub height 66m
Tilt angle .... 6°
Rotational speed (variable) 15-44rpm
Cone angle .... 3°

Performance
Rated power output 1,500kW
Operational wind speeds:
                cut in 4m/s
                rated 13.3m/s
                cut out 27m/s
Max. tip speed 138m/s

Power Control
Type .... yaw of the rotor/nacelle
Actuation .... 2 hydraulic motors

Rotor brake
Mechanical brake:
Type .... disc
Position .... low speed shaft
Actuation .... hydraulic
Air brake:
Type .... yawing

Rotor Blades
Material .... glass fibre/epoxy
Airfoil section .. NACA 230 xx series

Hub
Type .... teetered
Material .... cast steel

Gearbox
Type .... 2 x epicyclic
Ratio 1:32.57

Generator
Type .... synchronous
Rated power 1,500kW
Rated voltage 1,200V
Speed range (nominal) 1,433rpm

Yaw system
Actuation .... hydraulic
Yaw rate .... 2-8°/s

Nacelle
Base frame ...... welded steel bedplate
Cover .... glass fibre/polyester

Tower
Type .... tubular steel
Diameter (Top/Base) 3.0m

Masses
Blade (each) 6.5t
Hub 11t
Nacelle 61t
Towerhead 110t
Tower 130t
Total 240t

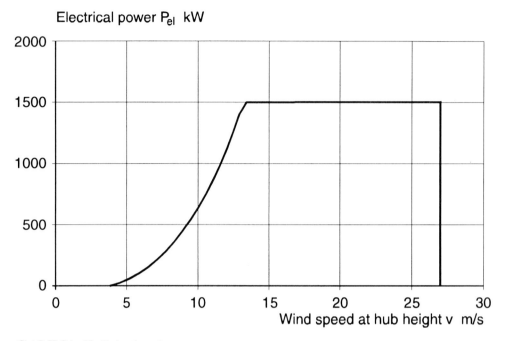

GAMMA 60 Calculated power curve

GAMMA 60 Manufacturing of the nacelle

GAMMA 60 Transport of the nacelle

GAMMA 60 Erection on the test site at Alta Nurra, Sardinia

First Operational Results

The GAMMA 60 1.5MW prototype was installed at Alta Nurra in April 1992. The turbine was purchased by ENEL within the cooperation agreement with ENEA. In December 1994, the GAMMA 60 achieved the production target of 1,000MWh set by ENEL as a condition for the completion of the commissioning phase. ENEL then continued the long-term evaluation of the wind turbine.

The GAMMA 60 prototype produced about 640MWh in 1994, achieving its 1,000MWh commissioning target on 23 December 1994. After installation in April 1992, an initial period of testing and adjusting followed. The experimental operation of the GAMMA 60 started in April 1993 with a limited power output of 750kW which gradually increased up to the rated value of 1,500kW. Apart from several improvements of performance and safety as usual for prototypes the GAMMA 60 operates as foreseen.

The industrialised version of the GAMMA 60 with a rated power of 2,000kW is under development.

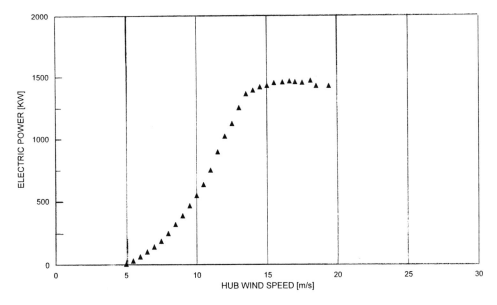

GAMMA 60 Measured power curve

NedWind NW50

General Description

The NedWind NW50 is one of the two prototypes of the commercial NW Megawatt-series. The turbines of this class are designed with a rotor diameter from 53 to 55m and a rated power of 1,000kW. Two units (2 x NW53) have been tested since 1994 and a third unit (NW55) will be installed in 1996. The NedWind NW53 is a two bladed active stall regulated turbine. The design is based on the NedWind NW40 (500kW) which is commercially used in large numbers.

The **rotor blades** of the NedWind NW50 series are approximately 25m long and of glass fibre reinforced polyester. The blades are manufactured by an automated technique which imitates the hand lay-up method. The same blades on blade extenders are used for the NW55 with a rotor diameter of 55m.

Turbines of the NedWind NW50 series have full span blade pitch actuation, i.e. the entire blade length is pitched which in turn influences the aerodynamic characteristics of the rotor. Six applications for blade pitching are:

- as an aerodynamic brake for safety reasons
- to optimize efficiency, especially at wind speeds below rated wind speeds
- to enable the induction of blade stall for power regulation
- to enable smooth grid connection by controlling the rotor acceleration
- to reduce the aerodynamic noise emissions at low wind speeds
- to enable parking of the rotor in horizontal position.

A special feature of the NedWind NW53 is the **power control by "active stall"**. As the power level, at which the blades stall, varies according to the turbulence intensity, the air density variations and the blade soiling, it is necessary to compensate for these variations by means of an adjustment mechanism. Blade pitching is well suited to perform these adjustments. It is also possible to design a turbine with very high power at stall and to reduce the excess power at high wind speeds by pitching the blades towards negative or stall. The advantage is that the rated wind speed can be achieved earlier, which results in a steeper power curve.

The **drive train** of the turbine is modular. In the NedWind NW50 series, four identical units have been developed: the adapter flange, connecting these four units to the gearbox, a mechanical brake, a flexible coupling and a 250kW generator. These same units are used in the NedWind NW30 series (250kW) and the NedWind NW40 series (500kW).

In order to minimize mechanical noise, the entire drive train of the megawatt turbine is mounted on elastomeric elements. The gearbox and main shaft with a length of more than 5m form one single unit. The hub end of the main shaft is supported by a self-aligning roller bearing (the main bearing) which supports the rotor weight. The main shaft is also supported inside the gearbox by two additional bearings. The gearbox is supported by two large elastomeric mounts and the main bearing is supported by four mounts. Thus, the drive train is mounted on three points. Due to the large distance between the mounts, the loads are small and rubber support elements can be used.

The **electric power conversion system** is also modular. A four-generator-system has the advantage that at partial load not all of the generators have to be connected to the grid at the same time. This increases the efficiency and improves the factor of the generated power compared to the use of only one generator with 1,000kW. The generators are connected to the grid via a thyristor controlled "soft-starter" which limits torque spikes to about one time of the rated torque (of the generator).

The dual speed machines, type NW55, in Moerdijk are fitted with double wound generators to be able to operate at 2rpm.

The NW53, as installed in Medemblik, has a **yaw drive system** consisting of a bull gear driven by pinions from two yaw motors. This system also has a yaw brake consisting of three brake calipers which can be applied or released as the case may be. The NW55 turbine, will be fitted with this yaw drive system.

The first NedWind megawatt turbine was installed in Spijk (NL) in Febuary 1994. The NW50 turbine was first connected to the grid on 23 March 1994 and has since then been measured and tested. The experience with the turbine in Spijk has given NedWind the opportunity to make adjustments to their second NW53 turbine, before installation in Medemblik (NL) on 23 May 1995. The Medemblik turbine is the world's first turbine of this dimension to be commercially operated from the moment of commissioning. The next project will be the installation of a wind farm near Moerdijk (NL), comprising four turbines, type NW55, to be installed in March 1996.

NedWind NW50 Sites of the first wind turbines in Spijk and Medemblik, The Netherlands

NedWind NW50 Prototype on the test site at Spijk, The Netherlands

WEGA Large Wind Turbines 99

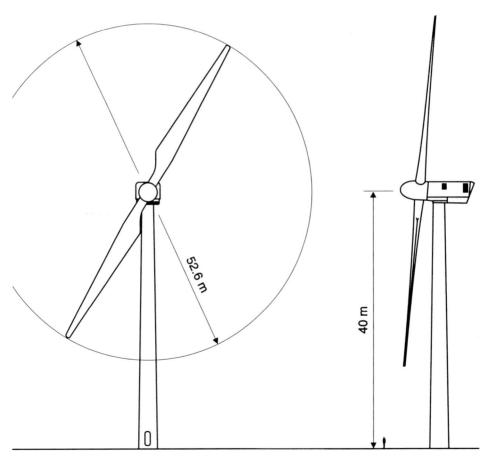

NedWind NW50 Side and front view, 1:500

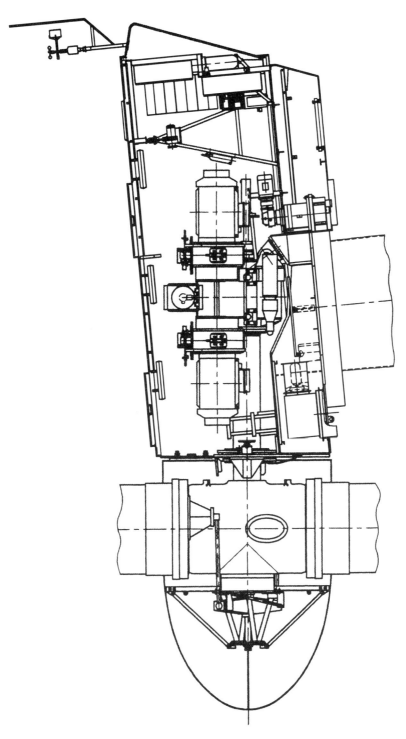

NedWind NW50 Cross section of the nacelle

Main Data of the NedWind NW50 Prototype

Rotor
Number of blades 2
Orientation .... upwind
Diameter 52.6m
Swept area .... 2,173m^2
Hub height 40m
Tilt angle .... 5°
Rotational speeds (fixed) ....
16.4/24.6rpm

Performance
Rated power output 1,000kW
Operational wind speeds:
               cut in 4m/s
               rated 14m/s
               cut out 20m/s
Design tip speed ratio 8
Max. tip speed 68m/s

Power Control
Type .... active stall
Actuation .... hydraulic

Rotor brake
Mechanical brake:
Type .... disc
Position .... at 4 high speed shafts
Actuation ... hydraulic./spring applied
Air brake:
Type .... full span blade pitch
Actuation .... hydraulic

Rotor Blades
Material .... glass fibre/polyester
Airfoil section NACA 636&646series

Hub
Type .... rigid
Material .... steel

Gearbox
Type .... 3 x parallel
Ratio 1:62

Generator
Type .... 4 x induction
Rated power 1,000kW
Rated voltage 400/690V
Speed 1,016/1,525rpm

Yaw system
Actuation .... hydraulic

Nacelle
Base frame ...... welded steel bedplate
Cover .... glass fibre/polyester

Tower
Type .... tubular steel

Masses
Blade (each) 5.2t
Rotor 21t
Nacelle 56t
Towerhead 77t
Tower 49t
Total 114t

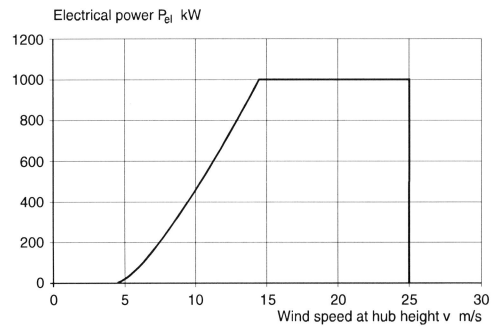

NedWind NW53 Calculated power curve

NedWind NW53 Rotor shaft arrangement

NedWind NW53 Gearbox and electric generators

NedWind NW53 Nacelle and rotor on the site, Medemblik

NedWind NW50 Erection on the test site at Spijk, The Netherlands

First Operational Results

Two units of the NW50/NW53 have been operating since March 1994 (Spijk) and since May 1995 (Medemblik) respectively. Essential operational experience and measurements have been acquired from the two machines. In 1994 the turbine NW53 was certified according to the Dutch standard NEN 6096/2.

Power Performance

The power curve of the NW50 has been measured in accordance with the IEA-standards and the Dutch NEN 6096/2.

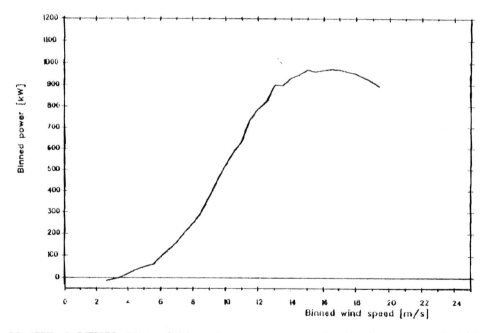

NedWind NW50 Plot of binned power curve, air density corrected with average value 1.24kg/m³

Noise Emission

The noise emission is an important parameter for the acceptability of a wind turbine. The NW50 in Spijk has a noise level of 103dB(A) at a wind speed of 8m/s measured 10m above ground, which is the minimum for a megawatt wind turbine. The noise level dependens mainly on the wind speed. At low wind speeds it is possible to reduce the noise level considerably by providing the turbine with a dual operating speed capability and by operating the turbine at low rpm. By doing so, the noise level can be reduced to 90dB(A), which is very low.

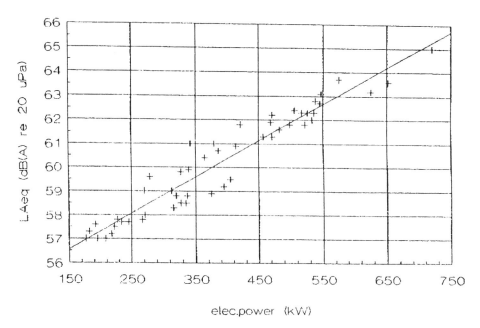

NedWind NW50 (Spijk) Measured levels at 24rpm, downwind corrected 60m

Nordtank NTK 1500

General Description

The overall concept of the NTK 1500 wind turbine originates from the „Danish concept": Upwind stall regulated, three bladed and fixed rotor speed for parallel grid operation.

The turbine is so to speak an upscale of the 500kW Nordtank turbine. However, the large turbine differs from the 500kW turbine in many respects regarding design: Three point suspension of main shaft/gear unit, two-split of the high speed shaft in the gear, two generators and finally a completely new look, improving the visual impact. The nacelle has been designed in collaboration with the well-known company „Jacob Jensen Design". During one year, several suggestions have been evaluated in respect to aesthetics, economy and technology. The final solution reflects the design philosophy of Nordtank turbines with an aerodynamic look in order to improve the visual impact in the landscape.

The **rotor blades** are based on the experience gained with the LM 17 blade. This blade has the well-known NACA profile type, working at a low noise level. The glass fibre polyester blades have been developed together with LM Glasfiber, Denmark. The rotor works with a tip speed of only 60m/s which is low compared to previous turbines and has a major effect on noise reduction.

The **hub** is spherical and thus simple to manufacture and with a good exploitation of material through the strength of the double-curved shell.

The design of the **drive train** has the following advantages:

- The three point suspension of the main shaft and gearbox minimizes alignment problems and is simplified in comparison to the four point suspensions of smaller turbines.

- One single main bearing on the main shaft reduces the technical complexity. The other main bearing is included in the gearbox where the greasing and lubricating surroundings are optimized.

- Because of the two output shafts on the gearbox, two separate brake discs can be applied. This increases the safety level of the mechanical braking system. The gear has been designed and developed in collaboration with Flender, Germany.

The **electric system** of the turbine includes two induction generators. The smaller size of the generators allows the possibility of variable speed by means of frequency control at lower costs. Only one generator is speed variable up to approximately 750kW. This generator has been developed and manufactured by Siemens, Germany. It is connected to the grid by soft-connection (thyristor controlled) reducing peak loads into the grid.

For smaller turbines, Nordtank manufactures the lower part of the **nacelle** in one single curved steel shell. This construction stiffens the machine and, moreover, includes the nacelle cover. For large wind turbines this solution is not the optimum. Therefore, a hollow box profile ensures more strength compared to the higher amount of material.

The design of the **yaw system** features the following advantages:

- As it is one separate module, easy service and replacement are ensured.
- Low friction ball bearing reduces the necessary yaw motor capacity.
- The bigger diameter of the tower top reduces the loads on the yaw motors.

The maximum diameter of the **tower** is constrained to 4.2m in view of transportation. Moreover, the double conic tower developed by Jacob Jensen Design gives the impression that the tower is connected firmly to the nacelle. Other designs with thinner tower tops give the impression of the nacelle balancing on the tower. The transformer is placed in the top of the tower in order to reduce energy losses in cables.

Regarding erection and maintenance of the turbine there are some important aspects:

- The nacelle can be hoisted in several smaller parts, thus reducing the necessary crane capacity.

- In case of damage, the generator and/or blade can be replaced by an internal crane.

- A separate elevator is provided for inspection and cleaning of the blades.

The prototype was erected on the wind turbine test field near Tjæreborg, Denmark in two days and nights (18 - 21 August 1995). The first connection to the grid took place on 7 September 1995 at a mean wind speed of 14 - 15m/s. The turbine produced 1.5MW within the first 10 min. in grid connected operation.

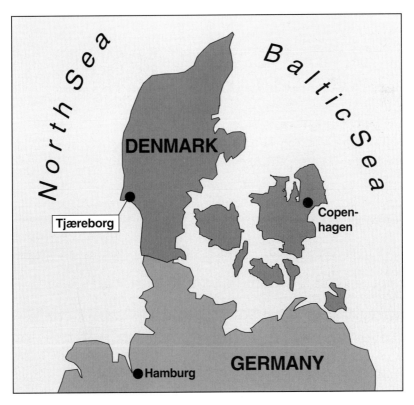

Nordtank NTK 1500 Site of the prototype at Tjæreborg, Denmark

Nordtank NTK 1500 Prototype at Tjæreborg, Denmark

WEGA Large Wind Turbines 111

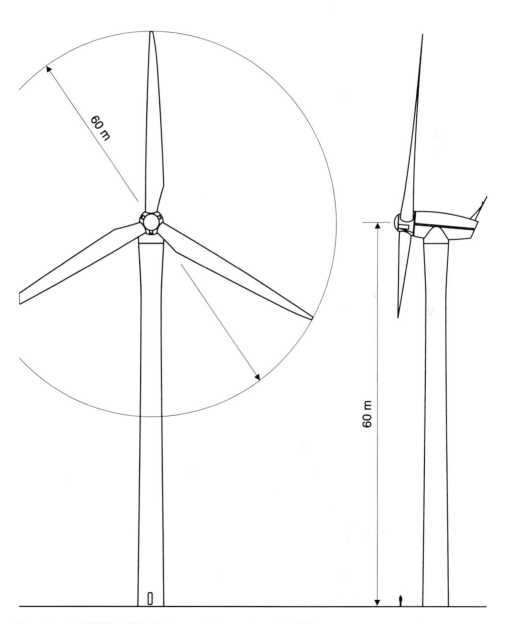

Nordtank NTK 1500 Side and front view, 1:500

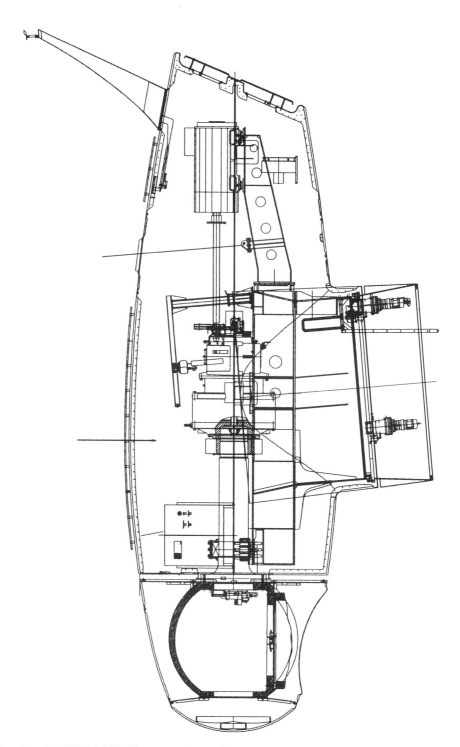

Nordtank NTK 1500 Cross section of the nacelle

Main Data of the Nordtank NTK 1500/60 Prototype

Rotor
Number of blades 3
Orientation .... upwind
Diameter 60m
Swept area .... 2,827m^2
Hub height 60m
Tilt angle .... 5°
Rotational speed (fixed) 19.2rpm

Performance
Rated power output 1,500kW
Operational wind speeds:
        cut in 4m/s
        rated 16m/s
        cut out 25m/s
Max. tip speed 60m/s

Power Control
Type .... stall

Rotor brake
Mechanical brake:
Type .... disc
Position .... at 2 high speed shafts
Actuation .... hydraulic/spring applied
Air brake:
Type .... tip brakes
Actuation .... centrifugal released/
.... hydraulic

Rotor Blades
Material .... glass fibre/polyester
Airfoil section ... NACA 63xxx series

Hub
Type .... rigid
Material .... cast steel

Gearbox
Type .... 1x epicyclic; 1 x parallel
Ratio 1:79

Generator
Type .... 2 x induction
Rated power 1,500kW
Rated voltage 690V
Speed (nominal) 1,500rpm

Yaw system
Actuation .... hydraulic

Nacelle
Base frame .... welded steel shell
Cover .... glass fibre/polyester

Tower
Type .... tubular steel

Masses
Blade (each) 5t
Towerhead 98t
Tower 95t
Total 193t

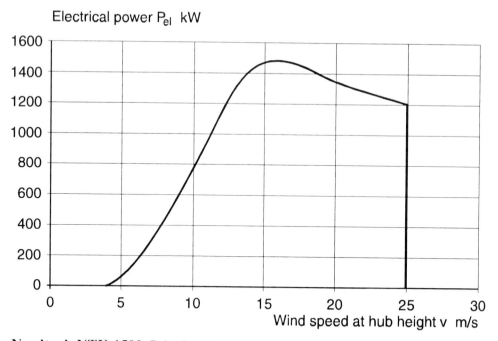

Nordtank NTK 1500 Calculated power curve

Nordtank NTK 1500 Gearbox on the test bench at Flender, Düsseldorf

Nordtank NTK 1500 Electric generator, Siemens test bench in Nürnberg

Nordtank NTK 1500 Rotor hub, Balle/Denmark

Nordtank NTK 1500 Nacelle and hub in the factory, Balle/Denmark

Nordtank NTK 1500 Mounting of the rotor

Nordtank NTK 1500
Erection on the test site at Tjæreborg, Denmark

First Operational Results

On 18 October 1995 RISO made the turbine operate in automatic mode after the blade test. During a blade test the flapwise and edgewise deflections were measured while applying a shear force on the measured blade (pulling from the tower). In fact, the turbine could have operated in automatic mode three weeks earlier, but the blade test had to be performed to meet RISO's demands.

At the end of September 1995 the turbine operated during a storm with gusts up to 32m/s. The turbine operated at nominal power for nearly 6 hours. On the basis of these findings, it was decided that the blades had to be pitched because the stall power was too high.

No server problems have been encountered during the operation of the prototype. In November 1995 the tip brakes were dismounted, because one of them could not regain its operating position (blade angle). The tips were tested and repaired by the blade manufacturer. The turbine was out of operation for nearly two weeks.

At the beginning of January 1996 further adjustments of the wind direction sensors took place. Afterwards the turbine operated correctly in the wind.

At the end of June 1996 the turbine has been operating grid connected for approximately 3,200 hours and has produced about 1,935MWh. The maximum availability reached 98% in June 1996.

Two noise measurements have been performed which were well below the 104dB, which is the criteria on the Tjæreborg site. The measurement indicates a level of approximately 102 dB(A) pressure level, no pure tones were observed.

Several measurements have been gathered for different situations and some adjustments to the measuring system have been made. The measuring system now operates ideally.

4
Scientific Measurement and Evaluation Project

Under the aspects of R&D it is the intention of the Commision to gather the general technological aspects of this wind turbine development and evaluation. To ensure that the results and the underlying measurements will be performed and presented on a comparable standard form, more specific guidelines will be distributed to the contractors later.

The Commission will focus on the dissemination and publication of the results of the general technological aspects. They will be compiled in agreement with the manufacturers of the wind turbines.

The basis of the outlined structure of the Scientific Measurement and Evaluation Project (SMEP) is the former measurement and evaluation programme for the large wind turbines of the WEGA-I /1/ programme. Whereas the WEGA-I measurement programme emphasized mainly the computer-based standardized data acquisition and exchange, the forthcoming programme will be focusing on two targets:

- scientific data acquisition and evaluation
- technical evaluation with respect to the general aspects
  of the wind turbine design verifications.

These two issues are both of equal importance to the development of the technology.

Monitoring System and Parameters to be measured

The monitoring system to be installed comprises a data acquisition system, sensors on the wind turbine and a meteorological mast. The list of parameters to be measured are:

Rotor blades

- blade bending moments, flapwise and edgewise
- blade pitch angle
- rotor azimuth angle

Nacelle and drive train

- low speed shaft torque and bending moment
- rotor speed
- rotor thrust
- nacelle position
- generator power
- electrical parameters relevant for grid interference
- self-consuming power of the WEC

Tower

- tower base bending moment
- tower top torsion moment

Wind

- wind speed at five heights up to hub height
- wind direction
- atmospheric pressure
- atmospheric temperature

Overall system

- essential eigenfrequencies of the main components
- noise emission parameters according to IEA recommendations

Monitoring Programme

The data shall be collected for the following load cases and operational modes:

Parked Rotor

- low and high wind speeds
- different blade pitch angle (if possible)

Start up

- below rated wind speed
- above rated wind speed

Steady Operation

- grid connected below rated wind speed (partial load)
- grid connected above rated wind speed (full load)
- different rotational speeds (if possible)

Normal Shut down

- below rated wind speed
- above rated wind speed

Emergency Shut down

In addition the contractor is free to carry out the measurement programme according to his own requirements and procedures. He only has to make sure that the requested minimum results will be reported to the Commission considering the guidelines for standardisation.

Expected Results

Functional Characteristics

Description of the functional behaviour considering the various operational modes (operational sequences).

Functional diagram of the safety system including a failure mode and effect analysis (description).

Power Curve

Cross and net power curve versus wind speed at hub height, based on IEA recommendations.

Noise Emission

Results of noise measurements according to IEA recommendations (2nd edition, 1988).

Power Control Quality

Time history of power output at low and high wind speeds

Yaw Quality

Comparison of nacelle azimuth angle and wind direction (time history and statistics)

Electrical Parameter of Grid Interference

Harmonics, reactive power consumption and voltage variations at various operational status.

Main Loads

- blade bending moments (flatwise and edgewise) (load spectra, mean values)
- main shaft torque (load spectrum, mean values)
- main shaft bending moment (load spectrum, mean value)
- yaw moment (load spectrum)

Resonance Diagram

including main subsystem and system eigenfrequencies

Time Histories of the Mentioned Loads for a Typical Operational Sequence

Operational Statistics

regular reports (every 3 months) including:

- operating hours (grid connected, downtime)
- energy production
- availability
- fault statistics

Wind Data
- monthly mean wind speed at different heights
- annual wind speed distribution
- wind direction distribution
- average wind shear exponent
- measured maximum wind speeds

Design Verification Report

By means of the predicted and measured main loads and functional characteristics a design verification report has to be elaborated. It has to consider the mentioned main loads and functional characteristics.

Standardisation of Data Acquisition and Presentation

To ensure comparability of the achieved results, a package of templates and guidelines will be worked out and provided to the contractor. However, this will be kept on a minimum level and as far as possible in the most simple form. The objective is to obtain comparable results and not to standardize the measurement procedures. This will be elaborated on the basis of the experience of the WEGA-I measurement project.

Time Schedule and Reporting

The monitoring and data evaluation will consider the general experience that the complete function of the computerized data acquisition system can only be achieved after a longer period. Therefore the measurement programme must be carried out in two phases:

Phase 1
- preliminary functional characteristics
- power curve
- noise emission
- power control quality
- electrical parameters of grid interference
- winddata

Phase 2
- yaw quality
- main loads
- resonance diagram
- operational statistics
- wind data

Following this strategy it will be ensured that in the case of technical problems and delays with the data acquisition system or with the turbine test operation very essential information can be obtained on a time basis.

The foreseen time period for the measurement and data evaluation programme is 12 months test operation. This anticipated an acceptable level of the availability of the wind turbine and of the data acquisition system.

The reporting of the results to the Commission has to comply with the following time schedule:

Regular reports:	Operational statistics (quarterly)
Results of Phase 1:	6 months after beginning of test operation
Results of Phase 2:	3 months after completion of Phase 1
Results of Phase 3:	6 months after completion of Phase 1

Organisation and Participants

ELSAMPROJEKT A/S, Denmark is the coordinator of the project. To assist ELSAMPROJEKT A/S in the evaluation work, a **"Central Evaluation Group"** consisting of: - besides ELSAMPROJEKT A/S - ETAPLAN GmbH, Germany; H. J. M. Beurskens (ECN), The Netherlands and F. Avia Aranda (IER-Ciemat), Spain has been formed.

ETAPLAN will compile the design and synopsis report outlining the main achievements of the projects of interest to the public.

The associated contractors responsible for performing the measurements of the wind turbines are:

Bonus A/S	: Bonus 750
Enercon GmbH	: E-66
Vattenfall AB	: Nordic 1000
Vestas Wind Systems A/S	: V63-1.5MW
Wind Energy Group Ltd	: WEG 600kW
Deutsches Windenergie-Institut (DEWI)	: AEOLUS II
Flygtekniska Forsöksanstalten (FFA)	: NÄSUDDEN II
WEST Spa	: GAMMA 60
NedWind Rhenen bv	: NW 53
Nordtank Energy Group A/S	: NTK1500

5
Further European Wind Turbines in the Megawatt Class

HSW 1000

General Description

The HSW 1000 is a three bladed, pitch-controlled horizontal axis wind turbine with an upwind rotor. The rotor diameter is 54m and the rated power 1,000kW. The hub height of the prototype is 55m. The HSW 1000 has been developed, tested and optimized during a two-year test-operation of a prototype with a nominal power of 750kW.

The **rotor blades** are made of glass fibre epoxy material. Connection to the rotor hub is achieved through a two-part modular cast iron flange. The blade bearings are four point contact bearings. Power regulation in the wind speed range 14 - 25m/s is achieved by hydraulic blade-pitch adjustment, so that nominal power of 1,000kW is delivered in this range.

The **rotor hub** is made of nodular cast iron and is designed as a spherical shell with three bolting plates for the blade bearing, one bolting plate for the rotor shaft flange and an access hole for installation and maintenance of the blade adjustment mechanism.

The **mechanical drive train** consists of a short main shaft, supported by two spherical roll bearings which are all integrated into one housing connected to the gearbox. This design allows a compact structure, common lubrication with the gearbox and also reduces the loadings on the bearings, as the weight of the gearbox is not supported by the bearings as in the case of a shaft mounted gearbox. The gearbox is a two-stage planetary gear with an additional spur gear stage. The gearbox has been designed for a continuous performance of 1,000kW and has a reduction ratio of i=63.06. The gearbox and rotor bearings are located in a common housing and are lubricated together. The gearbox is equipped with circulating oil lubrication and an oil cooler. The complete unit is bolted to the machine support in such a way that noise insulation is achieved.

The connection to the electric generator on the high speed side of the gear is cardan coupled with an intermediate shaft. An additional fluid coupling on the generator shaft balances the extremely small generator slip, and also serves to reduce torque peaks. In addition, the coupling operates as a safety coupling in case of a short-circuit of the generator. The rotor brake is located on the high speed side.

The **nacelle** comprises a bedplate (welded steel frame) and a GRP cover. A lifting crane is located in the nacelle. Tools and materials can be lifted into the nacelle through a small door located next to the generator. The nacelle is lined with sound-absorbant material. The ventilation-air flows through noise attenuation equipment. Consideration was given to noise abatement during the design of all noise generating equipment like the generator and the gearbox.

The **yaw control** is based on two opposing electric yaw motors. The yaw bearing is designed as a backlash-free four point contact bearing with external teething. The nacelle is arrested by spring-loaded brake calipers.

The **electrical system** includes a pole changeable induction generator (4/6 poled) with two speeds of 1,511/1,008rpm. The voltage is 690V. The ventilator has been designed as a low-noise type. The protective class of the generator is IP 54.

The control system is based on commercially available PLC controllers and a central unit. Remote control of the system per telephone modem is possible. Errors can thus be reported to a central monitoring station, if required.

Grid connection of the wind turbine is being realized at the level of the medium-high voltage system. The transformer and the switchboard can be installed in the tower or in a separate building.

The **tower** is a tubular steel design. To enable easy transport to the site, it is divided into three sections which are then bolted together. The diameter of the tower is 3.33m, the wall thickness varies between 12mm at the top and 26mm at the bottom. The tower is equipped with a ladder and free-hanging power cables.

The first prototype of the HSW 750 was connected to the grid on 6 August 1993. At the end of 1995 the first HSW 1000 turbines were erected at Bosbüll, Germany.

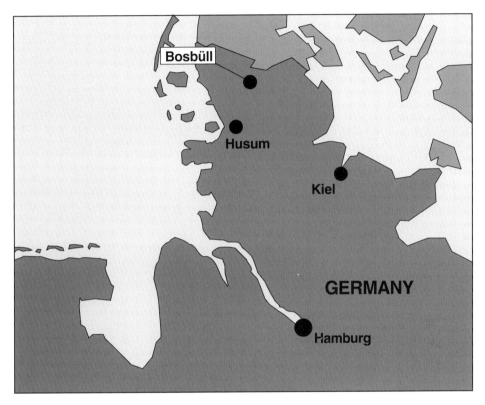

HSW 1000 Site of the prototype at Bosbüll, Germany

HSW 1000 Prototype at Bosbüll, Germany

WEGA Large Wind Turbines 133

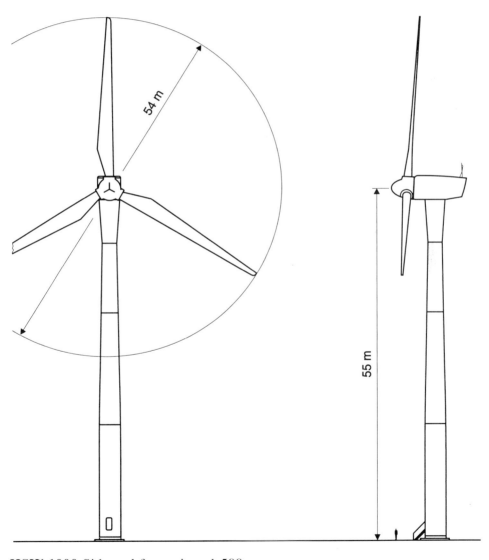

HSW 1000 Side and front view, 1:500

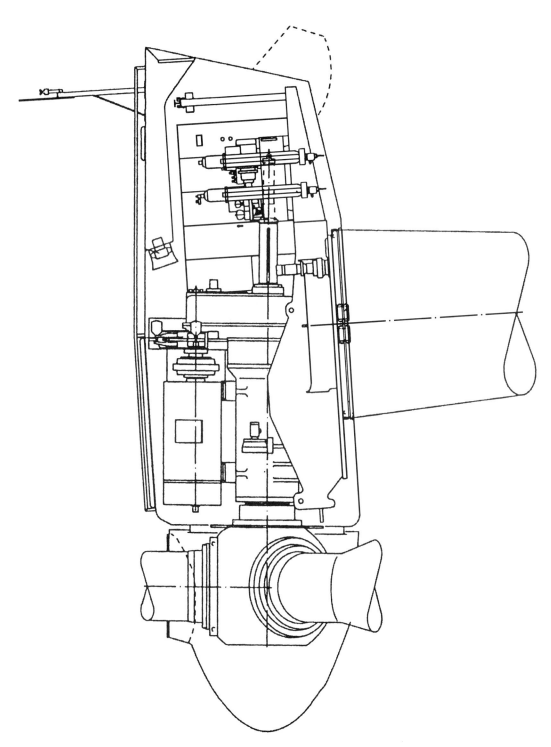

HSW 1000 Cross section of the nacelle

Main Data of the HSW 1000 Prototype

Rotor
Number of blades 3
Orientation .... upwind
Diameter 54m
Swept area .... 2,290m^2
Hub height 55m
Tilt angle .... 4°
Rotational sp. (fixed) ... 16.2/24.4rpm

Performance
Rated power output 1,000kW
Operational wind speeds:
        cut in 4m/s
        rated 13.5m/s
        cut out 28m/s
Survival wind speed 69m/s
Max. tip speed 69m/s

Power Control
Type .... full span blade pitch
Actuation .... hydraulic

Rotor brake
Mechanical brake:
Type .... disc
Position .... high speed shaft
Actuation ..... spring loaded/hydraulic
Air brake:
Type .... blade feathering

Rotor Blades
Material .... glass fibre/epoxy
Airfoil section .... NACA 44xx

Hub
Type .... rigid
Material .... cast steel

Gearbox
Type .... 1x epicyclic; 1 x parallel
Ratio 1:63.5

Generator
Type .... induction, 6/4 pole
Rated power 250/1,000kW
Rated voltage 690V
Speeds 1,008/1,511rpm

Yaw system
Actuation .... electrical

Nacelle
Base frame ...... welded steel bedplate
Cover .... glass fibre/polyester

Tower
Type .... tubular steel
Diameter (Base) 3.33m

Masses
Blade (each) 3.5t
Towerhead 73t
Tower 88t
Total 161t

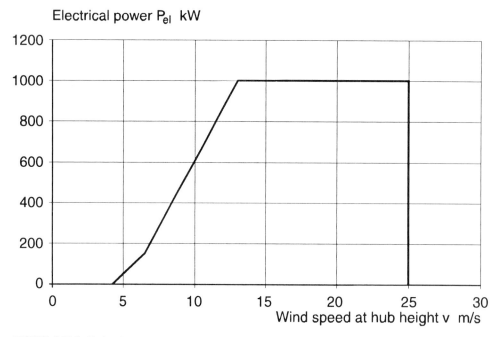

HSW 1000 Calculated power curve

Nordex N52

General Description

The Nordex N52 is a three bladed, stall controlled wind turbine. The upwind rotor has a diameter of 52.6m and a rated power output of 1,000kW. The first prototype (N52/800kW) has been tested since December 1994.

The **rotor blades**, supplied by LM (LM 24), are made of glass fibre polyester and are fitted with an integrated air brake system (blade tips). The blades are especially designed for stall operation. The tip brakes are actuated hydraulically. The blades are connected to the rotor hub via an extender giving a rotor diameter of 52.6m.

The **rotor hub** is cast steel and contains the hydraulic actuators for the blade tips.

The **mechanical drive train** is composed of several cast steel parts. The rotor shaft consists of a hollow load carrying part supporting the rotor weight. The driving shaft in the hollow support structure transmits only the torque to the gearbox. The gearbox is a Flender-design: epicyclic in the first stage, parallel in the 2nd and 3rd stages. On the high speed side of the gearbox a safety clutch and the rotor brake are placed between the gearbox and the generator.

The drive train has been designed and manufactured with special attention to noise insulation. The entire mechanical drive train is suspended by rubber elements. Between the rotor hub/electric generator and the shafts elastic couplings guarantee noise insulation.

The **nacelle** consists of a welded steel bedplate and a glass fibre cover structure. The GFRP structure includes lightning protection and a noise reduced cooling system for the generator and lubrication system.

The **yaw system** is driven by hydraulic motors and fail-safe brake calipers.

The **electric system** of the wind turbine is based on a pole-changeable induction generator (750rpm, 200kW // 1,500rpm; 1,000kW). The generator is fitted with a closed loop cooling system.

Operation is controlled by two computers. The first one is placed in the nacelle, the second one in the tower base. They are interconnected by an optical control cable. The main functions are continued by an uninterrupted power supply for a period of 10 minutes. The overall safety system also features redundant brake release and functional characteristics.

The **tower** of the prototype is a tubular steel tower composed of three segments (33t each). The tower base diameter is 4.04m. The wind turbine is also available with a 70m lattice tower.

The first prototype of the Nordex N52 (1,000kW) was connected to the grid in November 1995 in Borgholz.

Nordex N52 Site of the prototype at Borgholz, Germany

Nordex N52 Prototype at Borgholz, Germany

140   WEGA Large Wind Turbines

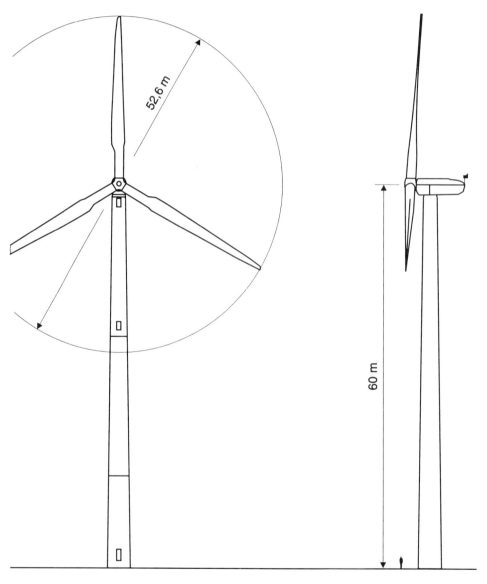

Nordex N52 Side and front view, 1:500

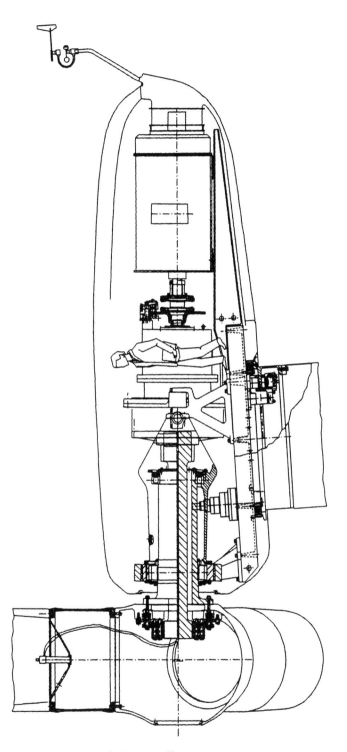

Nordex N52 Cross section of the nacelle

Main Data of the Nordex N52 Prototype

Rotor
Number of blades 3
Orientation .... upwind
Diameter 52.6m
Swept area .... 2,173m^2
Hub height 60m
Tilt angle .... 3°
Rotational speeds (fixed) 16.1/22rpm

Performance
Rated power output 1,000kW
Operational wind speeds:
                 cut in 3m/s
                 rated 16m/s
                 cut out 25m/s
Survival wind speed 55.8m/s
Max. tip speed 61m/s

Power Control
Type .... stall

Rotor brake
Mechanical brake:
Type .... disc
Position .... high speed shaft
Actuation ..... spring loaded/hydraulic
Air brake:
Type .... blade tips
Actuation .... centrifugal/hydraulic

Rotor Blades
Material .... glass fibre/polyester
Airfoil section LM 24 NACA 63-4xx
.... FFA-W3-2xx

Hub
Type .... rigid
Material .... cast steel

Gearbox
Type .... 1x epicyclic; 1 x spur
Ratio 1:45

Generator
Type .... induction, 8/6 pole
Rated power 200/1,000kW
Rated voltage 690V
Speed 750/1,000rpm

Yaw system
Actuation .... hydraulic
Yaw rate .... 0.5-0.6°/s

Nacelle
Base frame ...... welded steel bedplate
Cover .... glass fibre/polyester

Tower
Type .... tubular steel

Masses
Blade (each, incl. extender) 4.6t
Towerhead 74t
Tower 90t
Total 164t

WEGA Large Wind Turbines 143

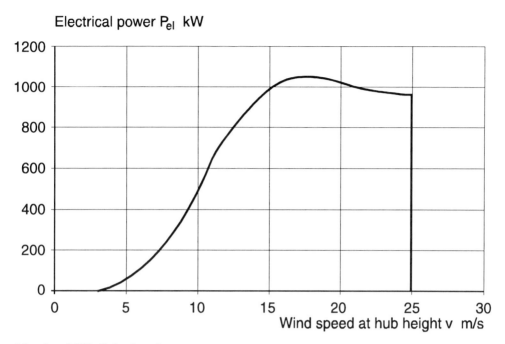

Nordex N52 Calculated power curve

Nordex N52 Erection of the tower

Nordex N52 Mounting of the nacelle

Nordex N52 Erection of the prototype at Borgholz, Germany

Tacke TW 1.5

General Description

The Tacke TW 1.5 is a three bladed, pitch controlled wind turbine. The rotor has a diameter of 65m and is located upwind from the tower. The rated power output is 1.500kW. The prototype has a hub height of 67m.

The **rotor blades** are of a special lightweight design and made of glass fibre and carbon fibre epoxy material.

The blades are connected to the **rotor hub** via four point contact roller bearings. The hub is cast steel. Pitch control of the blades is performed by individual electric motors for each blade.

The **mechanical drive train** of the Tacke TW 1.5 is integrated into the gearbox. The bearing case, attached to the gearbox includes rotor shaft and rotor bearings. The shaft is only loaded with the torque, whereas the rotor weight is supported by the casing of the bearings and the gearbox. The gearbox is a three stage epicyclic design.

The **nacelle** is composed by a compact load carrying bedplate and an aluminum cover. The cover is noise insulated by noise damping material.

The **yaw motion** is actuated by electric yaw motors which are controlled by power electronics. The brake calipers are fail-safe.

The **electric system** of the wind turbine is based on an induction generator. It features a slip-ring rotor and an inverter. By controlling the slip power, the generator can be operated speed variable between rated speed (20rpm) and 14rpm. The cooling system is a closed loop design.

The **safety-system** of the wind turbine operates by feathering the rotor blades. One blade feathering, independent from the other blades, can stop the rotor. The mechanical brake on the high speed side of the gearbox is used as parking brake.

The **tower** is a tubular steel design composed of four sections. It is fitted with internal ladders and the power cables.

The first prototype of the Tacke TW 1.5 was erected in March 1996. The connection to the grid took place on 17 April 1996.

Tacke TW 1.5 Site of the prototype at Emden, Germany

Tacke TW 1.5 Prototype at Emden, Germany

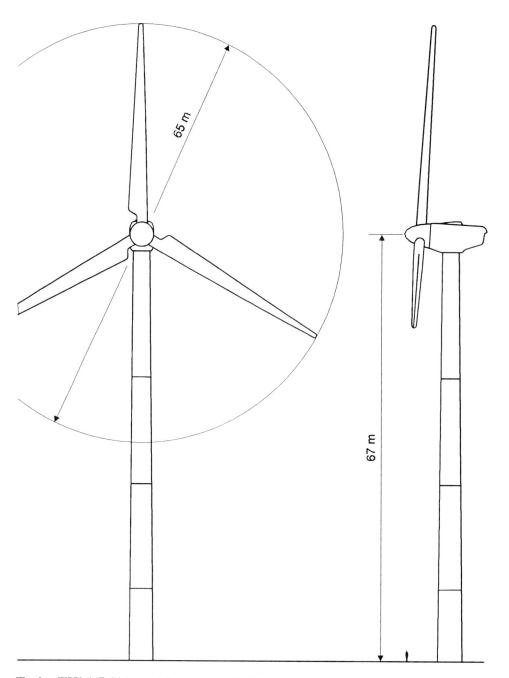

Tacke TW 1.5 Side and front view, 1:500

WEGA Large Wind Turbines 151

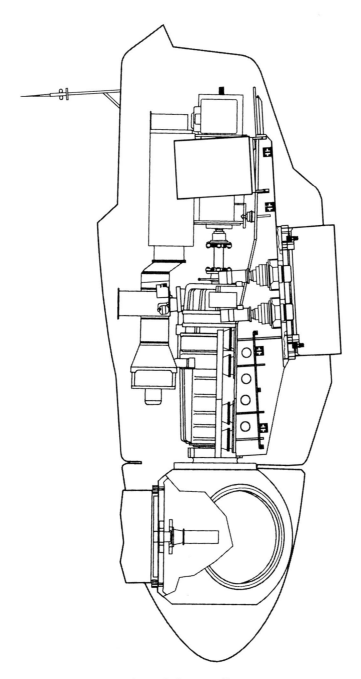

Tacke TW 1.5 Cross section of the nacelle

Main Data of the Tacke TW 1.5 Prototype

Rotor
Number of blades 3
Orientation .... upwind
Diameter 65m
Swept area .... 3,318m^2
Hub height 67m
Rotational speed (variable) 14-20rpm

Performance
Rated power output 1,500kW
Operational   cut in 4m/s
wind speeds: rated 13m/s
               cut out 25m/s
Survival wind speed 65.1m/s
Max. tip speed 68m/s

Power Control
Type .... full span blade pitch
Actuation .... hydraulic

Rotor brake
Mechanical brake:
Type .... disc
Position .... high speed shaft
Actuation .... hydraulic
Air brake:
Type .... blade feathering

Rotor Blades
Material ...... glass/carbon fibre epoxy

Hub
Type .... rigid
Material .... cast steel

Gearbox
Type .... 3 x epicyclic; 1 x parallel
Ratio 1:78

Generator
Type .... induction
Rated power 1,500kW
Rated voltage 500V
Speed 1,560rpm

Yaw system
Actuation .... electrical

Nacelle
Base frame ...... welded steel bedplate
Cover .... aluminium

Tower
Type .... tubular steel

Masses
Blade (each) 3.6t
Towerhead 97t
Tower 140t
Total 237t

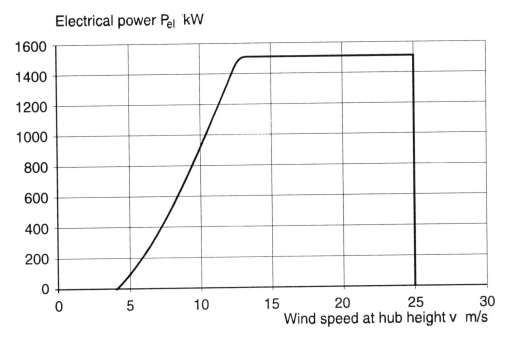

Tacke TW 1.5 Calculated power curve

Tacke TW 1.5
Tower and nacelle

Tacke TW 1.5 Assembly of the rotor

Tacke TW 1.5 Mounting of the rotor

References

/1/ E. Hau, J. Langenbrinck, W. Palz: 'WEGA Large Wind Turbines', Springer Verlag, 1993

/2/ E. Hau: 'Windkraftanlagen', 2. Auflage, Springer Verlag, 1995

/3/ E. Hau, R. Harrison, H. Snel: 'Design and Costs of Large Wind Turbines - Study on the next generation of large wind turbines', Commission of the European Communities, JOUR-0011-D (AM), to be published in 1996

/4/ P. Christiansen, P. Friis: 'WEGA-II Large Wind Turbines Scientific Evaluation Project', JOULE II, Contractors Meeting, Glasgow, 8 - 10 February 1995

The Communities research and development information service
CORDIS

A vital part of your programme's dissemination strategy

CORDIS is the information service set up under the VALUE programme to give quick and easy access to information on European Community research programmes. It is available free-of-charge online via the European Commission host organization (ECHO), and now also on a newly released CD-ROM.

CORDIS offers the European R&D community:

— a comprehensive up-to-date view of EC R&TD activities, through a set of databases and related services,
— quick and easy access to information on EC research programmes and results,
— a continuously evolving Commission service tailored to the needs of the research community and industry,
— full user support, including documentation, training and the CORDIS help desk.

The CORDIS Databases are:

**R&TD-programmes – R&TD-projects – R&TD-partners – R&TD-results
R&TD-publications – R&TD-comdocuments – R&TD-acronyms – R&TD-news**

Make sure your programme gains the maximum benefit from CORDIS

— Inform the CORDIS unit of your programme initiatives,
— contribute information regularly to CORDIS databases such as R&TD-news, R&TD-publications and R&TD-programmes,
— use CORDIS databases, such as R&TD-partners, in the implementation of your programme,
— consult CORDIS for up-to-date information on other programmes relevant to your activities,
— inform your programme participants about CORDIS and the importance of their contribution to the service as well as the benefits which they will derive from it,
— contribute to the evolution of CORDIS by sending your comments on the service to the CORDIS Unit.

For more information about contributing to CORDIS,
contact the DG XIII CORDIS Unit

Brussels
Ms I. Vounakis
Tel. +(32) 2 299 0464
Fax +(32) 2 299 0467

Luxembourg
M. B. Niessen
Tel. +(352) 4301 33638
Fax +(352) 4301 34989

To register for online access to CORDIS, contact:

ECHO Customer Service
BP 2373
L-1023 Luxembourg
Tel. +(352) 3498 1240
Fax +(352) 3498 1248

If you are already an ECHO user, please mention your customer number.

European Commission

EUR 16902 — WEGA II large wind turbines. Intermediate design report on the projects

Luxembourg: Office for Official Publications of the European Communities

1996 — vi, 157 pp. — 17.6 x 25.0 cm

ISBN 92-827-8903-9

Price (excluding VAT in Luxembourg): ECU 18.50